Topics in Torsion Theory

Paul E. Bland

MATHEMATICAL
RESEARCH

Volume 103

Topics in Torsion Theory

Paul E. Bland

Berlin · Weinheim · New York · Chichester
Brisbane · Singapore · Toronto

Author:
Prof. Paul E. Bland, Eastern Kentucky University, Richmond, Kentucky, USA

lst edition

Die Deutsche Bibliothek – CIP-Einheitsaufnahme

Bland, Paul E.:
Topics in torsion theory / Paul E. Bland. - 1. ed. - Berlin ; Weinheim ; New York ;
Chichester ; Brisbane ; Singapore ; Toronto : Wiley-VCH, 1998
(Mathematical research ; Vol. 103)
ISBN 3-527-40131-8

ISSN 0138-3019

Printed on non-acid paper.
The paper used corresponds to both the U. S. standard ANSI Z.39.48 – 1984
and the European standard ISO TC 46.

Printing and bookbinding: GAM Media GmbH, Berlin

Printed in the Federal Republic of Germany

WILEY-VCH Verlag Berlin GmbH
Mühlenstraße 33–34
D-13187 Berlin
Federal Republic of Germany

PREFACE

Torsion Theory provides an umbrella under which many classical properties of rings and modules can be recast in a more general form. The purpose of these notes is to furnish a brief introduction to torsion theory and to study properties of rings and modules in this setting. The goal has been to develop selected material on rings and modules in a torsion theoretical setting so that classical results can be recovered when an appropriate torsion theory is chosen. Hopefully, this work will (1) provide the reader with the foundation necessary to investigate research topics in torsion theory and (2) show how torsion theory relates to rings and modules. It is assumed that the reader has a knowledge of ring and module theory, at least at the level of a first year graduate course in abstract algebra which places some emphasis on this area.

The functors Ext and Tor from homological algebra are used either in comments or in proof but neither is used extensively. Category terminology is employed for the useful convenience of its language, although categorical arguments and abstractions have been avoided. The reader who has the necessary background in rings and modules, but who is unfamiliar with category theory, should have little difficulty in reading these notes. Moreover, the background information from homological algebra necessary to read this manuscript can be obtained quickly.

In preparing these notes, every effort was made to cite the relevant literature and to give credit to individual researchers. Hopefully, authors whose work is relevant to the material in these notes whom I have overlooked has been held to a minimum.

In a technical work of this length, mistakes may inadvertently occur. Errors should be credited to this author and not to the authors cited from the literature.

The preparation of these notes started while the author was on sabbatical leave visiting the University of Otago in Dunedin, New Zealand and was completed when the author returned to Eastern Kentucky University. The author would like to thank Eastern Kentucky University for providing the sabbatical leave and for providing additional support in various ways. I would also like to thank the faculty and staff of the University of Otago for making my stay there both pleasant and productive. In particular, I would like to thank John and Austina Clark who went out of their way to be very kind and helpful to me during my stay in New Zealand. Finally, I would like to thank my wife, Carole, who, during the preparation of the later portion of this manuscript, was willing to take on chores which we usually share. Had she not been willing to do this and had it not been for her patience and understanding, I certainly would not have had the time to devote to a work of this nature.

TABLE OF CONTENTS

Preface 5

Introduction 9

§1. Preliminaries on Torsion Theory 11

§2. The Jacobson Radical, Simple Modules and Chain Conditions Relative to Torsion Theory. Nakayama's Lemma 29

2.1 Simple Modules and Maximal Submodules Relative to a Torsion Theory 29

2.2 The Jacobson Radical and Radical Free Modules Relative to a Torsion Theory 33

2.3 Noetherian Modules Relative to a Torsion Theory 39

2.4 Artinian Modules Relative to a Torsion Theory 49

§3. Composition Series Relative to Torsion Theory. The Generalized Hopkins–Levitzki Theorem 55

§4. Injective and Projective Concepts and Torsion Theory. The Generalized Baer and Fuchs Conditions. Relatively Flat Modules 65

4.1 (Quasi–)Injective Modules Relative to a Torsion Theory 65

4.2 (Quasi–)Projective Modules Relative to a Torsion Theory 81

4.3 Flat Modules Relative to a Torsion Theory 88

§5. Covers and Hulls 101

5.1 (Quasi–)Injective Hulls Relative to a Torsion Theory 102

5.2 (Quasi–)Projective Covers Relative to a Torsion Theory 113

5.3 Torsionfree Covers 126

§6. Classical Properties of Rings Relative to a Torsion Theory 135

6.1 (Semi)Primitive, Simple and (Semi)Prime Rings Relative to a Torsion Theory 135

6.2 Semisimple Rings Relative to a Torsion Theory 143

6.3 Primitive Rings and Density Relative to a Torsion Theory 148

References 157

Index 159

INTRODUCTION

Throughout these notes R will denote an associative ring with identity and module or R–module will mean unitary right R–module. Mod–R and R–Mod will designate the category of unitary right and unitary left R–modules, respectively. The modifier "left" will be used when left R–modules are considered.

The preliminaries on torsion theory are established in §1 and with the beginning of §2 it is assumed that the torsion theory under consideration is hereditary. This assumption continues, unless stated otherwise, throughout the remainder of these notes. In §2, simple modules, maximal submodules and maximal right ideals are considered relative to a torsion theory and the relative Jacobson radical of a module is investigated. Modules with chain conditions are also studied relative to a torsion theory and a relative form of Nakayama's Lemma is given.

Modules with a relative composition series are investigated in §3 where it is shown that a module has a relative composition series if and only if it is a torsionfree module which is relatively Artinian and relatively Noetherian. The length of a relative composition series is defined and this is shown to be an invariant for each module which has such a series. §3 concludes with a generalized version of The Hopkins–Levitzki Theorem.

In §4, relative injective modules and related concepts are studied. An R–module M is injective if and only if each R–linear mapping from a right ideal of R to M can be extended linearly to R. This is known as Baer's Condition. Somewhat less well known is a result due to Fuchs which asserts that an R–module M is quasi–injective if and only if each R–linear mapping from a right ideal of R to M whose kernel contains the right annihilator of an element of M can be extended linearly to R. We refer to this as The Fuchs Condition. It is shown that both of these conditions can be generalized in the setting of an arbitrary torsion theory. The Generalized Baer Condition is well known while The Generalized Fuchs Condition is a more recent discovery. Projective, quasi–projective and flat modules are also studied with regard to how they relate to torsion theory.

In §5, injective hulls, quasi–injective hulls, projective covers and quasi–projective covers are investigated. The relative injective hull and the relative quasi–injective hull of a module are shown to exist universally and to be unique up to isomorphism. Relative projective and relative quasi–projective covers are also shown to be unique up to isomorphism when they exist. The assumption of universal existence of these covers is used to characterize those rings which are right perfect modulo their torsion ideal. Torsionfree covers are also investigated in §5 and it is shown that a torsion theory is universally covering (= every module has a torsionfree cover) whenever the Gabriel filter of right ideals of R associated with the torsion theory contains a cofinal subset of finitely generated right ideals. Although this condition is sufficient for a torsion theory to be universally covering, it is not known when, or even if, it is necessary. Intrinsic conditions in R which are both necessary and sufficient for a torsion theory to be universally covering are yet to be discovered. This work generalizes the work of Enochs who has shown that the usual torsion theory on Mod–Z is universally covering.

The work presented in §6 relates additional concepts in classical ring theory to torsion theory. We look at how (semi)simple, (semi)prime and (semi)primitive rings can be defined relative to a torsion theory so that classical results are recovered when the appropriate torsion theory is chosen. We believe this approach to investigating these classical properties of rings relative to torsion theory is new and we hope that these ideas will merit further investigation. In particular, rings which are right primitive with regard to a torsion theory are linked, via the work of Zelmanowitz, to m–density and monoform modules.

§1 PRELIMINARIES ON TORSION THEORY

Torsion theories were first introduced by S. E. Dickson in a paper entitled *A Torsion Theory for Abelian Categories* [10]. Dickson's work, published in the *Transactions of the American Mathematical Society* in 1966, was a generalization of the properties of the classes of torsion and torsionfree abelian groups to abelian categories. Since 1966 torsion theory has been an active area of research and it has become evident that torsion theories are ubiquitous in ring and module theory. We begin the study of selected properties of rings and modules in the setting of a hereditary torsion theory on Mod–R with terminology.

Definition 1.1.1 A class $\mathscr{C}$ of modules is said to be closed under: *homomorphic images* if $M \in \mathscr{C}$ implies that every homomorphic image of M is in $\mathscr{C}$; *submodules* if $M \in \mathscr{C}$ implies that every submodule of M is in $\mathscr{C}$; *direct sums (direct products)* if, whenever $\{ M_\alpha \}_{\alpha \in I}$ is a family of modules each of which is in $\mathscr{C}$, $\bigoplus_{\alpha \in I} M_\alpha \in \mathscr{C}$ ($\prod_{\alpha \in I} M_\alpha \in \mathscr{C}$); *extensions* if, whenever $0 \to L \to M \to N \to 0$ is a short exact sequence, L and N in $\mathscr{C}$ implies $M \in \mathscr{C}$; and *injective hulls* [11] (*projective covers* [3]) if the injective hull (projective cover, if such exists) of every module in $\mathscr{C}$ is also in $\mathscr{C}$.

Definition 1.1.2 A *torsion theory* τ on Mod–R is a pair $(\mathscr{T}, \mathscr{F})$ of classes of modules which satisfy the following conditions:

1. $\mathscr{T} \cap \mathscr{F} = 0$
2. If $M \to N \to 0$ is exact and $M \in \mathscr{T}$, then $N \in \mathscr{T}$.
3. If $0 \to M \to N$ is exact and $N \in \mathscr{F}$, then $M \in \mathscr{F}$.

4. For each module M there exist $L \in \mathcal{T}$ and $N \in \mathcal{F}$ such that

   $0 \to L \to M \to N \to 0$ is exact.

Modules in $\mathcal{T}$ are said to be τ–*torsion* and those in $\mathcal{F}$ are said to be τ–*torsionfree*. If $\mathcal{T}$ is closed under submodules, τ is said to be *hereditary*. Similarly, τ is called *cohereditary* when $\mathcal{F}$ is closed under homomorphic images. A class of modules $\mathcal{T}$ is said to be a (*hereditary*) *torsion class*, if there is a class of modules $\mathcal{F}$ such that $(\mathcal{T}, \mathcal{F})$ is a (hereditary) torsion theory on Mod–R. Similarly, a class of modules $\mathcal{F}$ is called (*cohereditary* and) *torsionfree*, if there is a class of modules $\mathcal{T}$ such that $(\mathcal{T}, \mathcal{F})$ is a (cohereditary) torsion theory on Mod–R. If $R \in \mathcal{F}$, then τ is a *faithful* torsion theory.

The following definition is useful for simplifying terminology.

Definition 1.1.3 Let $\mathcal{A}$ and $\mathcal{B}$ be non–empty classes of modules. If $\mathcal{A} = \{ M \mid Hom_R(M, N) = 0 \ \forall N \in \mathcal{B} \}$, then $\mathcal{A}$ is said to be the *left orthogonal compliment* of $\mathcal{B}$. If $\mathcal{B} = \{ N \mid Hom_R(M, N) = 0 \ \forall M \in \mathcal{A} \}$, then $\mathcal{B}$ is said to be the *right orthogonal compliment* of $\mathcal{A}$. We refer to $(\mathcal{A}, \mathcal{B})$ as a *complimentary pair* when $\mathcal{A}$ is the left orthogonal compliment of $\mathcal{B}$ and $\mathcal{B}$ is the right orthogonal compliment of $\mathcal{A}$.

Proposition 1.1.4 *τ is a torsion theory on* Mod–R *if and only if $(\mathcal{T}, \mathcal{F})$ is a complimentary pair.*

Proof. Suppose first that τ is a torsion theory on Mod–R and let $\mathcal{A}$ be the left orthogonal compliment of $\mathcal{F}$. If $M \in \mathcal{T}$ and $f \in Hom_R(M, N)$ where

$N \in \mathscr{F}$, it follows from 2 and 3 of Definition 1.1.2 and the exactness of $M \to f(M) \to 0$ and $0 \to f(M) \to N$ that $f(M) \in \mathscr{T} \cap \mathscr{F} = 0$. Thus, $\mathscr{T} \subseteq \mathscr{A}$. If $M \in \mathscr{A}$, because of 4 of Definition 1.1.2, we can find $L \in \mathscr{T}$ and $N \in \mathscr{F}$ such that $0 \to L \xrightarrow{f} M \xrightarrow{g} N \to 0$ is exact. But then $g = 0$ and so $f(L) = M$. Hence, by 2 of Definition 1.1.2, $M \in \mathscr{T}$ and so $\mathscr{A} \subseteq \mathscr{T}$. The proof that $\mathscr{F}$ is the right orthogonal compliment of $\mathscr{T}$ is similar.

Conversely, suppose that τ is a complimentary pair. We need to show that the classes $\mathscr{T}$ and $\mathscr{F}$ satisfy the four conditions of Definition 1.1.2.

1. If $0 \neq M \in \mathscr{T} \cap \mathscr{F}$, the identity map $M \to M$ shows that $\mathrm{Hom}_R(M, M) \neq 0$, a contradiction.
2. If $M \xrightarrow{f} N \to 0$ is exact with $M \in \mathscr{T}$ and $N \notin \mathscr{T}$, there is an $N' \in \mathscr{F}$ and $g \in \mathrm{Hom}_R(N, N')$ such that $g \neq 0$. But then $g \circ f \neq 0$ which contradicts the fact that τ is a complimentary pair.
3. If $0 \to M \xrightarrow{f} N$ is exact with $N \in \mathscr{F}$ and $M \notin \mathscr{F}$, there is an $N' \in \mathscr{T}$ and a $g \in \mathrm{Hom}_R(N', M)$ such that $g \neq 0$. But $f \circ g \neq 0$ and so τ would not be a complimentary pair.
4. Let $\mathscr{C}$ denote the collection of all τ–torsion submodules of an arbitrary module M. If $L = \sum_{T \in \mathscr{C}} T$, then L is a homomorphic image of $\bigoplus_{T \in \mathscr{C}} T$ and so since condition 2 of Definition 1.1.2 has been shown to hold, $L \in \mathscr{T}$ whenever $\bigoplus_{T \in \mathscr{C}} T \in \mathscr{T}$. But $\bigoplus_{T \in \mathscr{C}} T \in \mathscr{T}$, since $\mathrm{Hom}_R(\bigoplus_{T \in \mathscr{C}} T, N) \cong \prod_{T \in \mathscr{C}} \mathrm{Hom}_R(T, N) = 0$ for all $N \in \mathscr{F}$. Finally, we claim that $M/L \in \mathscr{F}$. If $M/L \notin \mathscr{F}$, there must be an $M' \in \mathscr{T}$ such that $\mathrm{Hom}_R(M', M/L) \neq 0$. If $0 \neq f \in \mathrm{Hom}_R(M', M/L)$, then $f(M') \in \mathscr{T}$. Let $L \subseteq X \subseteq M$ be such that $X/L \cong f(M')$. If $N \in \mathscr{F}$, it follows that $\mathrm{Hom}_R(X, N) = 0$ since $\mathrm{Hom}_R(L, N) = \mathrm{Hom}_R(X/L, N) = 0$ and $\mathrm{Hom}_R(X/L, N) \to \mathrm{Hom}_R(X, N) \to \mathrm{Hom}_R(L, N)$ is exact. But $\mathrm{Hom}_R(X, N) = 0$ for all $N \in \mathscr{F}$ tells us that

$X \in \mathscr{T}$. Therefore, $X \subseteq L$ in which case $L = X$. Since this contradicts the assumption that $f \neq 0$, $M/L \in \mathscr{F}$. Thus, we have found an $L \in \mathscr{T}$ and an $M/L \in \mathscr{F}$ such that $0 \to L \to M \to M/L \to 0$ is exact. Consequently, when τ is a complimentary pair, τ is a torsion theory. □

If τ is a torsion theory on Mod–R and $\mathscr{C}$ is the collection of all the τ–torsion submodules of M, we have seen in the proof of Proposition 1.1.4 that $t_\tau(M) = \sum_{T \in \mathscr{C}} T$ is that necessarily unique submodule of M such that $t_\tau(M) \in \mathscr{T}$ and $M/t_\tau(M) \in \mathscr{F}$. It now follows easily that M is τ–torsion if and only if $t_\tau(M) = M$ and τ–torsionfree if and only if $t_\tau(M) = 0$. Thus, we have $\mathscr{T} = \{ M \mid t_\tau(M) = M \}$ and $\mathscr{F} = \{ M \mid t_\tau(M) = 0 \}$. Note also that if $f : M \to N$ is R–linear, $f(t_\tau(M)) \subseteq t_\tau(N)$. From this it follows that $t_\tau(R)$ is an ideal of R. Indeed, if $r \in R$, the R–linear mapping $R \to R : x \to rx$ shows that $rt_\tau(R) \subseteq t_\tau(R)$.

Definition 1.1.5 If τ is a torsion theory on Mod–R, then $t_\tau(M)$ is referred to as the τ–*torsion submodule* of M and $t_\tau(R)$ is called the τ–*torsion ideal* of R. An R–module M is said to be *injective* if for every row exact diagram of R–modules and R–linear mappings of the form

$$\begin{array}{c} 0 \to L \to X \to N \to 0 \\ \downarrow \\ M \end{array}$$

can be completed commutatively by an R–linear mapping $X \to M$.
The *injective hull* [11] of an R–module M is an injective R–module E(M) together with an injective R–linear mapping $\varphi : M \to E(M)$ such that $\varphi(M)$

is an essential[1] submodule of M. It is well known that every R–module has an injective hull which is unique up to isomorphism.

Proposition 1.1.6 *If* τ *is a torsion theory on* Mod–R, *the following are equivalent:*

1. τ *is hereditary.*
2. $t_\tau(M) \cap N = t_\tau(N)$ *for every* R*–module* M *and each submodule* N *of* M.
3. $\mathcal{F}$ *is closed under injective hulls.*

Proof. $1 \Rightarrow 2$. Let N be a submodule of M and suppose that τ is hereditary. Since $t_\tau(M) \cap N$ is a submodule of $t_\tau(M)$, $t_\tau(M) \cap N$ is τ–torsion. Consequently, $t_\tau(M) \cap N \subseteq t_\tau(N)$ by virtue of how $t_\tau(N)$ is constructed. Since it is always the case that $t_\tau(N) \subseteq t_\tau(M) \cap N$, equality holds.

$2 \Rightarrow 3$. Suppose that $M \in \mathcal{F}$ and let E(M) denote the injective hull of M. Since $t_\tau(\,E(M)\,) \cap M = t_\tau(M) = 0$, we see that $t_\tau(\,E(M)\,) = 0$ since M is an essential submodule of E(M). Thus, $E(M) \in \mathcal{F}$.

$3 \Rightarrow 1$. Suppose that M′ is a submodule of a module $M \in \mathcal{T}$. If $M' \notin \mathcal{T}$, there must be an $N \in \mathcal{F}$ and a $0 \neq f \in \mathrm{Hom}_R(\,M', N\,)$. Since the row exact diagram

$$\begin{array}{ccccc} 0 & \rightarrow & M' & \rightarrow & M \\ & & f\downarrow & & \\ 0 & \rightarrow & N & \rightarrow & E(N) \end{array}$$

can be completed commutatively by a $0 \neq g \in \mathrm{Hom}_R(\,M, E(N)\,)$, this contradicts the fact that $\mathrm{Hom}_R(\,M, E(N)\,) = 0$. Hence, $M' \in \mathcal{T}$ and so τ is hereditary. □

[1] N is an essential submodule of M if $N \cap X \neq 0$ for every nonzero submodule X of M.

An R–module M is said to be *projective* if every row exact diagram of R–modules and R–linear mappings of the form

$$\begin{array}{c} M \\ \downarrow \\ 0 \to L \to X \to N \to 0 \end{array}$$

can be completed commutatively by an R–linear mapping $M \to X$.
A projective R–module together with an R–linear epimorphism$\varphi : P \to M$ is said to be a *projective cover* of M if $\ker \varphi$ is a small[2] submodule of P. Unlike injective hulls, projective covers may fail to exist. A ring R is said to be *right perfect* if every R–module has a *projective cover*. Bass [3] has characterized right perfect rights as being precisely those rings for which $R/J(R)$ is semisimple and $J(R)$ is *right T–nilpotent*. A subset I of R is right T–nilpotent, if for every sequence $x_1, x_2, \cdots$ in I, $x_n \cdots x_2 x_1 = 0$ for some integer $n \geq 1$.

Since injective hulls universally exist, every module in $\mathscr{F}$ has an injective hull. However, it may be the case that modules in $\mathscr{T}$ fail to have projective covers. This observation motivates the following definition.

Definition 1.1.7 A torsion theory τ is said to be *balanced* if every module in $\mathscr{T}$ has a projective cover.

Obviously, every torsion theory τ on Mod–R is balanced if and only if R is a right perfect ring. We have seen in Proposition 1.1.6 that a torsion theory

---

[2] N is said to be a small submodule of M if whenever X is a proper submodule of M, $N+X$ is also a proper submodule of M.

τ is hereditary if and only if $\mathscr{F}$ is closed under injective hulls. The following is the "dual" of this proposition.

Proposition 1.1.8 *If* τ *is a balanced torsion theory on* Mod–R, *then* τ *is cohereditary if and only if* $\mathscr{T}$ *is closed under projective covers.*

Proof. Suppose τ is cohereditary, $M \in \mathscr{T}$ and let $\varphi : P \to M$ be a projective cover of M. Since $P/\ker\varphi \cong M$, $P/\ker\varphi \in \mathscr{T}$ which tells us that $P/(\ker\varphi + t_\tau(P)) \in \mathscr{T}$. Now $P/(\ker\varphi + t_\tau(P))$ is a homomorphic image of $P/t_\tau(P)$ and so $P/(\ker\varphi + t_\tau(P)) \in \mathscr{F}$. Thus, $P/(\ker\varphi + t_\tau(P)) \in \mathscr{T} \cap \mathscr{F} = 0$ which indicates that $P = \ker\varphi + t_\tau(P)$. But $\ker\varphi$ is small in P and so $P = t_\tau(P)$. Therefore, $P \in \mathscr{T}$.

Conversely, suppose that $\mathscr{T}$ is closed under projective covers and consider any R–linear epimorphism $f : N \to N'$ where $N \in \mathscr{F}$ and $N' \neq 0$. If $N' \notin \mathscr{F}$, then for any $M \in \mathscr{T}$, $\mathrm{Hom}_R(M, N') \neq 0$. Let $g : M \to N'$ be R–linear and nonzero. Then $g^{-1}(N') \in \mathscr{T}$ and so has a projective cover $\varphi : P \to g^{-1}(N')$. If $h = g|_{g^{-1}(N')}$, then there is a nonzero R–linear mapping $k : P \to N$ such that $f \circ k = h \circ \varphi$. But this is impossible since $P \in \mathscr{T}$ and $N \in \mathscr{F}$. Hence, it must be that case that $N' \in \mathscr{F}$. □

A torsion theory has been shown to be a complimentary pair of classes of modules in Mod–R. It is often useful to know the properties which characterize torsion and torsionfree classes for a torsion theory on Mod–R.

Proposition 1.1.9 *The following hold for classes of modules* $\mathscr{T}$ *and* $\mathscr{F}$ *in* Mod–R.

1. *$\mathscr{T}$ is a torsion class for a torsion theory on* Mod–R *if and only if* $\mathscr{T}$ *is closed under homomorphic images, direct sums and extensions.*
2. *$\mathscr{F}$ is a torsionfree class for a torsion theory on* Mod–R *if and only* if $\mathscr{F}$ *is closed under submodules, isomorphic images, direct products and extensions.*

Proof. We only prove 2 since the proof of 1 is similar to the proof of Proposition 1.1.4. Suppose that $\mathscr{F}$ is a torsionfree class for a torsion theory τ on Mod–R. It follows immediately from Definition 1.1.2 of a torsion theory that $\mathscr{F}$ is closed under submodules and isomorphic images. Now let $\{ M_\alpha \}_{\alpha \in I}$ be a family of module each of which is in $\mathscr{F}$. If $M \in \mathscr{T}$, then $\mathrm{Hom}_R\left(M, \prod_{\alpha \in I} M_\alpha \right) \cong \prod_{\alpha \in I} \mathrm{Hom}_R(M, M_\alpha) = 0$ and so it follows that $\mathscr{F}$ is closed under direct products. Next, suppose $L \subseteq N$ is such that L and N/L are in $\mathscr{F}$. If $M \in \mathscr{T}$, then $\mathrm{Hom}_R (M, L) = \mathrm{Hom}_R(M, N/L) = 0$ and so the exactness of $0 \to \mathrm{Hom}_R(M, L) \to \mathrm{Hom}_R(M, N) \to \mathrm{Hom}_R(M, N/L)$ tells us that $\mathrm{Hom}_R(M, N) = 0$. Hence, $N \in \mathscr{F}$ and so $\mathscr{F}$ satisfies all the conditions in 2.

Conversely, suppose $\mathscr{F}$ is a class of modules which is closed under submodules, isomorphic images, direct products and extensions. Let $\mathscr{T}$ be the left orthogonal compliment of $\mathscr{F}$ and set $\tau = (\mathscr{T}, \mathscr{F})$. We claim that τ is a torsion theory on Mod–R. We show that the four conditions of Definition 1.2 are satisfied.

1. If $M \in \mathscr{T} \cap \mathscr{F}$, the identity map is in $\mathrm{Hom}_R (M, M) = 0$ and so $M = 0$.
2. Let $M \xrightarrow{f} N \to 0$ be exact with $M \in \mathscr{T}$. If $N \notin \mathscr{T}$, there is an $N' \in \mathscr{F}$ and a $0 \neq g \in \mathrm{Hom}_R(N, N')$. But then $0 \neq g \circ f \in \mathrm{Hom}_R(M, N')$, a contradiction. Hence, $N \in \mathscr{T}$.

3. Let $0 \to M \xrightarrow{f} N$ be exact where $N \in \mathscr{F}$. Since $f(M) \subseteq N$ and $M \cong f(M)$, it follows from the fact that $\mathscr{F}$ is closed under submodules and isomorphic images that $M \in \mathscr{F}$.
4. For an arbitrary module M, $0 \to t_\tau(M) \to M \to M/t_\tau(M) \to 0$ is exact with $t_\tau(M) \in \mathscr{T}$ and $M/t_\tau(M) \in \mathscr{F}$.

Thus, τ is a torsion theory on Mod–R as was to be shown. □

Definition 1.1.10 If τ is a torsion theory on Mod–R and M is an R–module, $F_\tau(M)$ will denote the set of all submodules N of M such that $M/N \in \mathscr{T}$. In this case, we will say that N is a τ*–dense submodule* of M. If N is τ–dense in M, we also say that M is a τ*–dense extension* of N. When τ is hereditary, $F_\tau(M)$ is usually referred to as the *filter of* τ*–dense submodules* of M. If $N \in F_\tau(M)$ is also an essential submodule of M, we say that N is τ*–essential* in M and that M is a τ*–essential extension* of N.

When τ is hereditary, the following proposition justifies calling $F_\tau(M)$ a filter.

Proposition 1.1.11 *When* τ *is a hereditary torsion theory on* Mod–R, *the following hold for any* R*–module* M.

1. *If* $L \in F_\tau(M)$ *and* N *is a submodule of* M *such that* $L \subseteq N$, *then* $N \in F_\tau(M)$.
2. *If* $L, N \in F_\tau(M)$, *then* $L \cap N \in F_\tau(M)$.

Proof. 1 follows immediately from the fact that $\mathscr{T}$ is closed under homomorphic images. For $L, N \in F_\tau(M)$, the map $M \to M/L \oplus M/N$: $m \to (m+L, m+N)$ has kernel $L \cap N$ and so $M/(L \cap N)$ embeds in

$M/L \oplus M/N \in \mathscr{T}$. Thus, $M/(L \cap N) \in \mathscr{T}$ since τ is hereditary. Hence, $L \cap N \in F_\tau(M)$ and so we have 2. □

Gabriel filters [19] are of particular interest since they are closely associated with hereditary torsion theories on Mod–R.

Definition 1.1.12 A non-empty collection $F(R)$ of right ideals of R is said to be a *Gabriel filter* if:

1. If $K \in F(R)$ and $x \in R$, $(K : x) = \{ r \in R \mid xr \in K \} \in F(R)$.
2. If $J \in F(R)$ and K is a right ideal of R such that $(K : x) \in F(R)$ for each $x \in J$, then $K \in F(R)$.

Proposition 1.1.13 *If* $F(R)$ *is a Gabriel filter of right ideals of* R, *then:*

1. *If* $J \in F(R)$ *and* $J \subseteq K$, *then* $K \in F(R)$.
2. *If* $J, K \in F(R)$, *then* $J \cap K \in F(R)$.
3. *If* $J, K \in F(R)$, *then* $JK \in F(R)$.

Proof. First note that $R \in F(R)$. This follows since if $K \in F(R)$ and $x \in K$, $R = (K : x) \in F(R)$ by 1 of Definition 1.1.12.

1. If $J \in F(R)$ and $J \subseteq K$, then for each $x \in J$, $(K : x) = R \in F(R)$. Hence, it follows from 2 of Definition 1.1.12 that $K \in F(R)$.

2. If $J, K \in F(R)$ and $x \in K$, then $(J \cap K : x) = (J : x) \in F(R)$. Thus, by 2 of Definition 1.12, $J \cap K \in F(R)$.

3. Note that if $J, K \in F(R)$ and $x \in J$, then $K \subseteq (JK : x)$. It now follows from 1 of this proposition and 2 of Definition 1.1.12 that $JK \in F(R)$. □

Notice that if $F(R)$ is a Gabriel filter of right ideals of R, then $F(R)$ is a subset of $\mathscr{P}(R)$, the power set of R. Hence, the collection of all such filters is a subset of $\mathscr{P}(\mathscr{P}(R))$ and so is a set. The following proposition shows there is a one–to–one correspondence between Gabriel filters and hereditary torsion theories on Mod–R. Hence, the collection of all hereditary torsion theories on Mod–R is also a set.

Proposition 1.1.14 *If* ψ *and* ϕ *are defined on the collection* $\mathscr{H}$ *of all hereditary torsion classes* $\mathscr{T}$ *in* Mod–R *and the collection* $\mathscr{G}$ *of all Gabriel filters* F(R) *of right ideals of* R, *respectively, by*

$$\mathscr{T} \overset{\psi}{\rightarrow} F(R) = \{ K \subseteq R \mid K \textit{ is a right ideal of } R \textit{ such that } R/K \in \mathscr{T} \} \textit{ and}$$

$$F(R) \overset{\phi}{\rightarrow} \mathscr{T} = \{ M \mid (0:m) \in F(R)\ \forall\, m \in M \},$$

then ψ *and* ϕ *are well defined functions which establish a one–to–one correspondence between* $\mathscr{H}$ *and* $\mathscr{G}$.

Proof. Suppose that $\mathscr{T}$ is a hereditary torsion class in Mod–R and let $F(R) = \{ K \subseteq R \mid R/K \in \mathscr{T} \}$. We will show that conditions 1 and 2 of Definition 1.1.12 hold for the right ideals of F(R).

1. Consider the monomorphism $R/(K:x) \rightarrow R/K$ defined by $r + (K:x) \rightarrow xr + K$ where $K \in F(R)$ and $x \in R$. Since $R/K \in \mathscr{T}$, it follows from the properties of a hereditary torsion class that $R/(K:x) \in \mathscr{T}$. Thus, $(K:x) \in F(R)$.

2. Let $J \in F(R)$ and suppose that K is a right ideal of R such that $(K:x) \in F(R)$ for each $x \in J$. If $x + (J \cap K) \in J/(J \cap K)$, then $(0 : x + (J \cap K)) = (J \cap K : x) = (K:x) \in F(R)$ and so $J/(J \cap K) \in \mathscr{T}$. But $(J + K)/K \cong J/(J \cap K)$ tells us that $(J + K)/K \in \mathscr{T}$. Next, we see that $R/(J + K) \in \mathscr{T}$ since the canonical map $R/J \rightarrow R/(J + K)$ is an epimorphism and $\mathscr{T}$ is closed under homomorphic images. Now the

exactness of the sequence $0 \to (J + K)/K \to R/K \to R/(J + K) \to 0$ leads to $R/K \in \mathcal{T}$ because $\mathcal{T}$ is closed under extensions. Thus, $K \in F(R)$. Consequently, we see that a hereditary torsion class $\mathcal{T}$ determines a Gabriel filter $F(R)$ of right ideals of R.

Conversely, suppose that $F(R)$ is a Gabriel filter of right ideals of R and let $\mathcal{T} = \{ M \mid (0 : m) \in F(R) \ \forall\, m \in M \}$. If $\mathcal{T}$ is to be a hereditary torsion class, we must prove that $\mathcal{T}$ is closed under each of the following:

Submodules: Let N be a submodule of $M \in \mathcal{T}$. If $n \in N$, clearly $(0 : n) \in F(R)$ and so $N \in \mathcal{T}$.

Homomorphic Images: Suppose $M \to N$ is an epimorphism and that $M \in \mathcal{T}$. If $m \to n$, then $(0 : m) \subseteq (0 : n)$ and because of 1 of Proposition 1.1.13, $(0 : n) \in F(R)$.

Direct Sums: Let $\{ M_\alpha \}_{\alpha \in I}$ be a family of modules each of which is in $\mathcal{T}$. If $(m_\alpha) \in \bigoplus_{\alpha \in I} M_\alpha$, then $(0 : (m_\alpha)) = \bigcap_{\alpha \in I} (0 : m_\alpha)$. Since $m_\alpha = 0$ for almost all $\alpha \in I$, $\bigcap_{\alpha \in I} (0 : m_\alpha)$ can be written as a finite intersection and so $(0 : (m_\alpha)) \in F(R)$. Therefore, $\bigoplus_{\alpha \in I} M_\alpha \in \mathcal{T}$.

Extensions: Let $0 \to L \to M \to M/L \to 0$ be exact with L and M/L in $\mathcal{T}$. If $m \in M$, then $m + L \in M/L$ and so $(L : m) = (0 : m + L)$ is in $F(R)$. If $r \in (L : m)$, then $mr \in L$ and so $((0 : m) : r) = (0 : mr)$ is also in $F(R)$. Hence, $((0 : m) : r) \in F(R)$ for each $r \in (L : m)$. But in view of 2 of Definition 1.1.12, we see that $(0 : m) \in F(R)$. Hence, $M \in \mathcal{T}$. The more general case for an exact sequence $0 \to L \to M \to N \to 0$ with L and N in $\mathcal{T}$ now follows through the use of isomorphisms. Thus, a Gabriel filter of right ideals in a ring R determines a torsion class of modules in Mod–R.

Finally, we must show that the mappings ψ and ϕ are inverses of each other and that these mappings are well defined. We will show that

$\phi \circ \psi = 1_{\mathscr{H}}$ and leave the proof of $\psi \circ \phi = 1_{\mathscr{G}}$ and the fact that ψ and ϕ are well defined to the diligent reader. Toward this end, let $\mathscr{T}$ be a torsion class in Mod–R. Then $\psi(\mathscr{T})$ is the Gabriel filter $F(R) = \{ K \subseteq R \mid R/K \in \mathscr{T} \}$ and $\phi(F(R)) = \mathscr{T}^*$ where $\mathscr{T}^* = \{ M \mid (0:m) \in F(R)\ \forall\, m \in M\}$. We will be finished, if we can show $\mathscr{T} = \mathscr{T}^*$. If $M \in \mathscr{T}$ and $m \in M$, then $R/(0:m) \cong mR \in \mathscr{T}$ since $\mathscr{T}$ is closed under submodules. Thus, $(0:m) \in F(R)$ and so $M \in \mathscr{T}^*$ by virtue of the definition of $\mathscr{T}^*$. Conversely, let $M \in \mathscr{T}^*$. Then for each $m \in M$, $(0:m) \in F(R)$ and so $mR \cong R/(0:m) \in \mathscr{T}$. Since $\mathscr{T}$ is closed under direct sums, $\bigoplus_{m \in M} mR \in \mathscr{T}$. But the canonical map $\bigoplus_{m \in M} mR \to M$ is an epimorphism and so $M \in \mathscr{T}$. Therefore, $\mathscr{T} = \mathscr{T}^*$. □

Hence, we see that there is a one–to–one correspondence between Gabriel filters $F(R)$ of right ideals of R and hereditary torsion theories τ on Mod–R. The Gabriel filter corresponding to the hereditary torsion theory τ will now be denoted by $F_\tau(R)$.

Corollary 1.1.15 *Let* $F_\tau(R)$ *be the Gabriel filter of right ideals of* R *corresponding to the hereditary torsion theory* τ *on* Mod–R. *Then for any module* M, $t_\tau(M) = \{ m \in M \mid (0:m) \in F_\tau(R) \}$.

Proof. Let $\mathscr{T}$ be the hereditary torsion class corresponding to $F_\tau(R)$. Since $t_\tau(M) = \sum_{T \in \mathscr{C}} T$ where $\mathscr{C}$ is the collection of all τ–torsion submodules of M, we know that $t_\tau(M) \in \mathscr{T}$. Thus, $(0:m) \in F_\tau(R)$ for all $m \in t_\tau(M)$ and so $t_\tau(M) \subseteq \{ m \in M \mid (0:m) \in F_\tau(R) \}$. For the reverse containment, let $m \in M$ be such that $(0:m) \in F_\tau(R)$. Then $mR \cong R/(0:m) \in \mathscr{T}$ and so $mR \subseteq t_\tau(M)$. Hence, $m \in t_\tau(M)$ and so we have equality. □

Definition 1.1.16 Let $\mathscr{C}$ be a nonempty class of modules. If $\mathscr{F}$ is a right orthogonal compliment of $\mathscr{C}$ and $\mathscr{T}$ is a left orthogonal compliment of $\mathscr{F}$, $(\mathscr{T}, \mathscr{F})$ is a torsion theory on Mod–R. It will be referred to as *the torsion theory generated by* $\mathscr{C}$. If $\mathscr{T}$ is the left orthogonal of $\mathscr{C}$ and $\mathscr{F}$ is the right orthogonal compliment of $\mathscr{T}$, then $(\mathscr{T}, \mathscr{F})$ is a torsion theory on Mod–R. In this case, $(\mathscr{T}, \mathscr{F})$ is *the torsion theory cogenerated by* $\mathscr{C}$.

The following two propositions show that given a hereditary torsion theory τ on Mod–R, there are modules $X \in \mathscr{T}$ and $Y \in \mathscr{F}$ such that $\{X\}$ and $\{Y\}$ generate and cogenerate τ, respectively. Moreover, Y can be taken to be an injective module.

Proposition 1.1.17 *Every hereditary torsion theory* τ *on* Mod–R *is cogenerated by a* τ*–torsionfree injective module.*

Proof. Let τ be a hereditary torsion theory on Mod–R with τ–torsionfree class $\mathscr{F}$. Suppose also that $\mathscr{S}$ is a complete set of representatives of the cyclic τ–torsionfree modules. If $Y = E\left(\prod_{X \in \mathscr{S}} X\right)$ is the injective hull of $\prod_{X \in \mathscr{S}} X$, then $Y \in \mathscr{F}$ since $\mathscr{F}$ is closed under direct products and injective hulls. We claim that $\{Y\}$ cogenerates τ. If $\mathscr{A} = \{ M \mid \mathrm{Hom}_R(M, Y) = 0 \}$, we immediately have $\mathscr{T} \subseteq \mathscr{A}$ and so let $M \in \mathscr{A}$. If $M \notin \mathscr{T}$, then there is an $N \in \mathscr{F}$ such that $\mathrm{Hom}_R(M, N) \neq 0$. Let $0 \neq g \in \mathrm{Hom}_R(M, N)$ and suppose that N' is a nonzero cyclic submodule of $g(M)$. Now N' is τ–torsionfree and so $N' \cong X$ for some $X \in \mathscr{S}$. If $M' = g^{-1}(N')$ and $f = g|_{M'}$, then $f : M' \to N'$ is nonzero and from this we obtain a nonzero homomorphism $f^* : M' \to X$. Hence, $\mathrm{Hom}_R(M', X) \neq 0$ and so it follows that $\mathrm{Hom}_R(M', Y) \neq 0$. Since Y is injective, the exactness of $0 \to M' \to M$ yields an exact sequence $\mathrm{Hom}_R(M, Y) \to \mathrm{Hom}_R(M', Y) \to 0$. But $M \in \mathcal{A}$ and so $\mathrm{Hom}_R(M, Y) = 0$. Hence, $\mathrm{Hom}_R(M', Y) = 0$ which is a contradiction. Thus, $M \in \mathcal{T}$ and therefore $\mathcal{A} = \mathcal{T}$. Consequently, $\{Y\}$ cogenerates τ. □

The cogenerating injective module Y is by no means unique. If we define two injective modules to be equivalent if each can be embedded in a direct product of copies of the other, this is an equivalence relation on the class of injective modules. It can be shown [20] that there is a one–to–one correspondence between the collection $\mathcal{E}$ of equivalence classes of injective modules and the collection $\mathcal{H}$ of hereditary torsion theories on Mod–R.

Proposition 1.1.18 *If τ is a hereditary torsion theory on* Mod–R, *then τ is generated by a direct sum of cyclic τ–torsion modules.*

Proof. (Sketch) Let $\mathcal{S}$ be a complete set of representatives of the isomorphism classes of the cyclic τ–torsion modules and suppose that $F_\tau(R)$ is the Gabriel filter associated with τ. If $F_\tau(R)^* = \{ K \in F_\tau(R) \mid R/K \in \mathcal{S} \}$, set $X = \bigoplus_{K \in F_\tau(R)^*} R/K$. Then X is τ–torsion and in a manner similar to the proof of Proposition 1.1.17, it can be shown that $\mathcal{F}$ is the right orthogonal compliment of $\{X\}$ so that $\{X\}$ generates τ. □

We conclude our remarks on the basic properties of torsion theories with the following definition followed by several examples.

Definition 1.1.19 If $\sigma = (\mathcal{T}_\sigma, \mathcal{F}_\sigma)$ and $\tau = (\mathcal{T}_\tau, \mathcal{F}_\tau)$ are torsion theories on Mod–R, we write $\sigma \leq \tau$ whenever $\mathcal{T}_\sigma \subseteq \mathcal{T}_\tau$ or, equivalently, when $\mathcal{F}_\tau \subseteq \mathcal{F}_\sigma$.

Observe that $\leq$ is a partial order on the collection of all torsion theories on Mod–R.

Examples 1.1.20

1. If G is an abelian group, an element $x \in G$ is said to be torsion if there is an integer $n \neq 0$ such that $xn = 0$. If $t(G)$ denotes the set of all torsion elements of G, then $t(G)$ is the torsion subgroup of G. $\tau = (\mathscr{T}, \mathscr{F})$ is a faithful hereditary torsion theory on Mod–Z where $\mathscr{T} = \{ G \mid t(G) = G \}$ and $\mathscr{F} = \{ G \mid t(G) = 0 \}$. The Gabriel filter $F_\tau(Z)$ contains every nonzero ideal of Z. This torsion theory on Mod–Z is the precursor of the definition of a torsion theory on Mod–R.

2. The hereditary torsion theory cogenerated by the injective hull of R is called the *Lambek torsion theory* [14, 15] on Mod–R. It will be denoted by τ_L. A right ideal K is in $F_L(R)$ if and only if the left annihilator of $(K : x)$ is zero for each $x \in R$. Thus, $K \in F_L(R)$ if and only if for all $x, y \in R$, $y \neq 0$, there is an $r \in R$ such that $xr \in K$ and $yr \neq 0$. The right ideals in $F_L(R)$ are said to be *Lambek dense* in R.

3. An element $m \in M$ is said to be a *singular element* of M if $(0 : m)$ is an essential right ideal of R. If $Z(M)$ is the collection of all singular elements of M, $Z(M)$ is a submodule of M called the *singular submodule* of M. For the ring R, $Z(R)$ is an ideal of R called the *right singular ideal* of R. A module M is said to be *singular* if $Z(M) = M$ and *nonsingular* if $Z(M) = 0$. The class $\mathscr{N}$ of all nonsingular modules is a torsionfree class for a hereditary torsion theory on Mod–R. This torsion theory is called the *Goldie torsion*

theory [22] on Mod–R and it will be denoted by τ_G. For any module M, $t_G(M) = \{ m \in M \mid m + Z(M) \in Z(M/Z(M)) \}$. If K is an essential right ideal of R, $K \in F_G(R)$. Since every Lambek dense right ideal of R is essential in R, it follows that $\tau_L \leq \tau_G$. When R is a nonsingular ring, a right ideal of R is essential if and only if it is Lambek dense. Consequently, for a nonsingular ring, $\tau_L = \tau_G$. For an arbitrary torsion theory on Mod–R, notice that if M is τ–torsionfree and N is any τ–dense submodule of M, then N is essential in M. To see this, suppose M is τ–torsionfree and that $m \in M \setminus N$. Then $(mR + N)/N \subseteq M/N \in \mathcal{T}$ and so $(mR + N)/N \in \mathcal{T}$. But $(mR + N)/N \cong mR/(mR \cap N)$ and so if $mR \cap N = 0$ we would have a contradiction since mR is τ–torsionfree. If the torsion theory on Mod–R is faithful, then R is τ–torsionfree. Hence, if $K \in F_\tau(R)$, then as we have seen K is essential in R and so $K \in F_G(R)$. Consequently, $\tau \leq \tau_G$ for every faithful torsion theory τ on Mod–R. Hence, τ_G is a upper bound for the faithful torsion theories on Mod–R. τ_G may not be faithful.

4. Suppose that R is a subring of S such that S is a flat left R–module, then $\mathcal{T} = \{ M \mid M \otimes_R S = 0 \}$ is a hereditary torsion class in Mod–R. The class $\mathcal{F} = \{ M \mid M \to M \otimes_R S : m \to m \otimes 1_S$ is a monomorphism $\}$ is the torsionfree class corresponding to $\mathcal{T}$. The Gabriel filter $F_\tau(R)$ corresponding to this torsion theory τ is the set of right ideals K of R such that $KS = S$. This follows from the fact that $R/K \otimes_R S \cong S/KS$.

5. Let τ be a hereditary torsion theory on Mod–R with Gabriel filter $F_\tau(R)$. A left R–module M will be called τ–*divisible* if $KM = M$ for each right ideal $K \in F_\tau(R)$. If $\mathcal{D} = \{ M \in R\text{–Mod} \mid N \otimes_R M = 0 \ \forall N \in \mathcal{T} \}$, we claim that $\mathcal{D}$ is the class of all τ–divisible left R–modules. If $K \in F_\tau(R)$ and $M \in \mathcal{D}$,

then $0 = R/K \otimes_R M \cong M/KM$ and so $KM = M$. Thus, M is τ–divisible. Conversely, if $KM = M$ for all $K \in F(R)$, let $n \otimes m$ be a generator of $N \otimes_R M$ where $N \in \mathscr{T}$. Since $n \in N$, there is a $K \in F_\tau(R)$ such that $nK = 0$. Now $KM = M$, so let $m_i \in M$ and $k_i \in K$ be such that $\sum_{i=1}^{q} k_i m_i = m$. Then $n \otimes m = n \otimes \sum_{i=1}^{q} k_i m_i = \sum_{i=1}^{q}(n \otimes k_i m_i) = \sum_{i=1}^{q}(nk_i \otimes m_i) = \sum_{i=1}^{q}(0 \otimes m_i) = 0$. Thus, $N \otimes_R M = 0$ and so $M \in \mathscr{D}$. Therefore, $\mathscr{D}$ is the class of all τ–divisible left R–modules. $\mathscr{D}$ is a torsion class on R–Mod but $\mathscr{D}$ may not be hereditary even though $\mathscr{T}$ is hereditary. If every module in $\mathscr{T}$ is flat, then $\mathscr{D}$ will be hereditary. If $t_{\mathscr{D}}(N)$ is the sum of the τ–divisible submodules of M, M will be called τ–*reduced* when $t_{\mathscr{D}}(M) = 0$. The pair $(\mathscr{D}, \mathscr{R})$ is a torsion theory on R–Mod where $\mathscr{R}$ is the collection of all reduced left R–modules. When τ is the torsion theory of torsion and torsionfree abelian groups of Example 1, $\mathscr{D}$ and $\mathscr{R}$ are the usual classes of divisible and reduced abelian groups.

6. Finally, there are two important torsion theories which are embedded in classical abstract algebra. Namely, $(\mathscr{T}, 0)$ and $(0, \mathscr{F})$. $(\mathscr{T}, 0)$ is the torsion theory in which every module is torsion and $(0, \mathscr{F})$ is the torsion theory in which every module is torsionfree. Each is hereditary with Gabriel filter F(R) which contain every right ideal of R and Gabriel filter F(R) which contains only the ring R, respectively. Many results about rings and modules relative to a torsion theory on Mod–R reduce to classical results in abstract algebra when τ is chosen to be either $(\mathscr{T}, 0)$ or $(0, \mathscr{F})$.

§2 THE JACOBSON RADICAL, SIMPLE MODULES AND CHAIN CONDITIONS RELATIVE TO TORSION THEORY NAKAYAMA'S LEMMA

Throughout the remainder of these notes we assume, unless stated otherwise, that τ *is a hereditary torsion theory on* Mod–R. $\mathscr{T}$ and $\mathscr{F}$ will denote the torsion and torsionfree classes associated with τ, respectively. As indicated before, the Gabriel filter associated with τ will be denoted by $F_\tau(R)$. If more that one torsion theory is involved, subscripts will make the distinction between torsion classes, between torsionfree classes and between Gabriel filters.

2.1 Simple Modules and Maximal Submodules Relative to a Torsion Theory

Definition 2.1.1 A nonzero R–module M is said to be τ–*simple* if $M \in \mathscr{F}$ and every nonzero submodule of M is in $F_\tau(M)$. A submodule N of M will be called τ–*maximal* if M/N is τ–simple.

It follows immediately from the definition that if M is τ–simple, then every proper homomorphic image of M is in $\mathscr{T}$. Likewise, any nonzero submodule of a τ–simple module is τ–simple. Obviously, a τ–maximal submodule of M must be a proper submodule of M.

Many authors refer to τ–simple modules as τ–cocritical modules and to τ–maximal submodules as τ–critical submodules. We depart from existing terminology in order to emphasize the connection between the results contained in these notes and classical results in abstract algebra. For instance, when $\tau = (0, \mathscr{F})$ is the torsion theory in which every R–module is torsionfree, any τ–simple module is simple and a τ–maximal submodule of M is a maximal submodule of M. It may be the case that τ–simple R–modules fail to exist. For example, if $\tau = (\mathscr{T}, 0)$ is the torsion theory in which every R–module is torsion, there are no τ–simple modules. Therefore, *in order to ensure that our discussion is not vacuous, we assume that the torsion theory* τ *is such that* τ*–simple modules exist.*

Proposition 2.1.2 *The ring* R *has* τ*–maximal right ideals.*

Proof. Since we are assuming that τ is such that τ–simple modules exist, let M be a τ–simple R–module. If $0 \neq m \in M$, then mR is a τ–simple R–module. But $R/(0:m) \cong mR$ and so $(0:m)$ is a τ–maximal right ideal of R. □

Definition 2.1.3 A submodule N of an R–module M will be called τ*–pure* in M, if $M/N \in \mathscr{F}$ When R is considered as an R–module, a right ideal will be called τ–pure, if it is τ–pure when viewed as an R–submodule of R. An ideal of a ring R will be called τ–pure, if it is τ–pure as a right ideal. $\mathscr{P}_\tau(M)$ will denote the set of all τ–pure submodules of M.

Proposition 2.1.4 *If* $\{ N_\alpha \}_{\alpha \in \Delta}$ *is a family of* τ*–pure submodules of an* R*–module* M, *then* $\cap_{\alpha \in \Delta} N_\alpha$ *is a* τ*–pure submodule of* M.

Proof. If $m + \cap_{\alpha \in \Delta} N_\alpha$ is a τ–torsion element of $M/\cap_{\alpha \in \Delta} N_\alpha$, then $\left(0:m + \cap_{\alpha \in \Delta} N_\alpha\right) \in F_\tau(R)$. But $\left(0:m + \cap_{\alpha \in \Delta} N_\alpha\right) \subseteq (0:m + N_\alpha)$

for each $\alpha \in \Delta$ and so $(0 : m + N_\alpha) \in F_\tau(R)$ for each $\alpha \in \Delta$. Thus, $m + N_\alpha \in t_\tau(M/N_\alpha) = 0$ for each $\alpha \in \Delta$. Hence, $m \in N_\alpha$ for each $\alpha \in \Delta$ and so $m + \cap_{\alpha \in \Delta} N_\alpha = 0$. Consequently, $M/\cap_{\alpha \in \Delta} N_\alpha \in \mathscr{F}$. □

Definition 2.1.5 If N is a submodule of M an R–module M, N^c will denote the intersection of all the τ–pure submodules of M which contain N. We refer to the τ–pure submodule N^c as the τ–*pure closure* of N in M.

Recall that the τ–torsion submodule $t_\tau(M)$ of an R–module M is the sum of all the τ–torsion submodules of M and so, if the submodules of M are ordered by inclusion, $t_\tau(M)$ is the largest τ–torsion submodule of M. The following proposition shows that $t_\tau(M)$ is, in the sense of containment, the smallest τ–pure submodule of M.

Proposition 2.1.6 *For any* R–*module* M, $t_\tau(M) = \cap_{N \in \mathscr{P}_\tau(M)} N$.

Proof. Clearly, $\cap_{N \in \mathscr{P}_\tau(M)} N \subseteq t_\tau(M)$, since $t_\tau(M)$ is a τ–pure submodule of M. Now suppose $N \in \mathscr{P}_\tau(M)$ and $m \in t_\tau(M)$. Then $(0 : m) \subseteq (0 : m + N)$ tells us that $m + N$ is a τ–torsion element of the τ–torsionfree module M/N Thus, $m + N = 0$ and so $m \in N$. Hence, $t_\tau(M)$ is contained in each τ–pure submodule of M. □

Note that the proposition above indicates that $t_\tau(M)$ is the τ–pure closure of 0 in M.

Proposition 2.1.7 *If* N *is a submodule of an* R–*module* M, *then* N is τ–*dense in* N^c *and, in fact,* $N^c/N = t_\tau(M/N)$. *Moreover,* $N^c = \{ m \in M \mid (N : m) \in F_\tau(R) \}$.

Proof. Suppose $\mathscr{C}$ is the collection of all the τ–pure submodules of M which contain N. By definition $N^c = \cap_{X \in \mathscr{C}} X$ and so $N^c/N = \left(\cap_{X \in \mathscr{C}} X \right) / N = \cap_{X \in \mathscr{C}} X/N = t_\tau(M/N)$. Finally, note that $m \in N^c$ if and only if $m + N \in N^c/N = t_\tau(M/N)$ if and only if $(N : m) \in F_\tau(R)$. □

Proposition 2.1.8 *N is a τ–maximal submodule of an R–module M if and only if N is maximal among the proper τ–pure submodules of M. Moreover, if $M \in \mathscr{F}$, then any minimal element (should it exist) in the set of nonzero τ–pure submodules of M is τ–simple.*

Proof. Let N be a τ–maximal submodule of M and suppose that $X \in \mathscr{P}_\tau(M)$ is such that $N \subset X \subseteq M$. Since the canonical map $M/N \to M/X$ is an epimorphism, $M/X \in \mathscr{T}$. Thus, $M/X \in \mathscr{T} \cap \mathscr{F} = 0$ and so $X = M$.

Conversely, suppose N is maximal among the proper τ–pure submodules of M. If M/N is not τ–simple, the fact that $M/N \in \mathscr{F}$ implies there must exist a nonzero submodule X/N of M/N such that $(M/N)/(X/N) \notin \mathscr{T}$. Hence, $X \neq M$. Next suppose $(M/N)/(X/N) \in \mathscr{F}$. Then $M/X \in \mathscr{F}$ and so $X \in \mathscr{P}_\tau(M)$. But $X/N \neq 0$ tells us that N is a proper submodule of X which contradicts the maximality of N. Thus, $(M/N)/(X/N)$ is neither a τ–torsion or a τ–torsionfree R–module and so $t_\tau((M/N)/(X/N)) \neq 0$ is a proper submodule of $(M/N)/(X/N)$. Let Y be a proper submodule of M properly containing N such that $(Y/N)/(X/N) = t_\tau((M/N)/(X/N))$. Then $M/Y \cong (M/N)/(Y/N) \cong [(M/N)/(X/N)]/[(Y/N)/(X/N)] \in \mathscr{F}$. Thus, $Y \in \mathscr{P}_\tau(M)$ which again contradicts the maximality of N. Hence, M/N must be τ–simple and so N is τ–maximal.

Finally, suppose $M \in \mathscr{F}$. Let N be minimal among the nonzero modules in $\mathscr{P}_\tau(M)$ and note that $N \in \mathscr{F}$. If N is not τ–simple, there is a nonzero submodule A of N such that $N/A \notin \mathscr{T}$. Now suppose that B where $A \subseteq B \subseteq N$ is such that $B/A = t_\tau(N/A)$ then, $N/B \cong (N/A)/(B/A) = (N/A)/(t_\tau(N/A)) \in \mathscr{F}$. Thus, $B \in \mathscr{P}_\tau(N)$. Since $\mathscr{F}$ is closed under extensions, $0 \to N/B \to M/B \to M/N \to 0$ shows $B \in \mathscr{P}_\tau(M)$. But this contradicts the minimality of N and so N must be τ–simple. □

2.2 The Jacobson Radical and Radical Free Modules Relative to a Torsion Theory

Definition 2.2.1 The τ–*radical* $J_\tau(M)$ of an R–module M is defined to be the intersection of all the τ–maximal submodules of M. If an R–module fails to have τ–maximal submodules, we set $J_\tau(M) = M$. If $J_\tau(M) = 0$, we say that M is a τ–*radical free module*. $J_\tau(R)$ will denote the intersection of all the τ–maximal right ideals of R.

Since $J_\tau(M)$ is the intersection of the τ–maximal submodules of M, $J_\tau(M) \in \mathscr{P}_\tau(M)$. It follows immediately from this observation and Proposition 2.1.6 that $t_\tau(M) \subseteq J_\tau(M)$. Consequently, any τ–radical free R–module is in $\mathscr{F}$. If M is τ–simple, 0 is a maximal τ–pure submodule of M and so $J_\tau(M) = 0$. Thus, any τ–simple module is τ–radical free. If the set of τ–maximal submodules of M is coincident with the set of maximal submodules of M, for example, when $\tau = (0, \mathscr{F})$, $J_\tau(M)$ coincides with the Jacobson radical J(M) of M [25]. In this case, a τ–radical free module is often referred to as being radical free or as being Jacobson semisimple.

We have seen that N is a τ–maximal submodule of M if and only if N is maximal in $\mathscr{P}_\tau(M)$. The following shows that a τ–maximal submodule of M is also maximal in another set of submodules of M.

Proposition 2.2.2 N *is a* τ*–maximal submodule* M *if and only if* N is a *maximal among the submodules of* M *which are not in* $F_\tau(M)$.

Proof. First, suppose that N is a τ–maximal submodule of M. Then $N \notin F_\tau(M)$. Let N′ be a proper submodule of M such that $N \subseteq N'$ and $N' \notin F_\tau(M)$. Now $M/N' \neq t_\tau(M/N') = N'^c/N'$ and so $N'^c \neq M$. Hence, N'^c is a proper τ–pure submodule of M and $N \subseteq N' \subseteq N'^c$. But N is maximal among the proper τ–pure submodules of M and so $N = N' = N'^c$.

Conversely, let N be maximal among the submodules of M which are not in $F_\tau(M)$. Then $M/N \neq t_\tau(M/N) = N^c/N$ and so $N \subseteq N^c$. But $N^c \in \mathscr{P}_\tau(M)$ and so $N^c \notin F_\tau(M)$. Consequently, $N = N^c$ and so $N \in \mathscr{P}_\tau(M)$. Finally, if $X \in \mathscr{P}_\tau(M)$ is such that $N \subseteq X \subset M$, then $X \notin F_\tau(M)$ and so, by assumption, N = X, Thus, N is a τ–maximal submodule of M. □

Definition 2.2.3 A submodule N of M is said to be τ–*small* in M, if there are no proper τ–pure submodules X of M such that $N^c + X \in F_\tau(M)$.

It is well know that the Jacobson radical J(M) of M contains every small submodule of M. In fact, J(M) is the sum of the small submodules of M. The following relates $J_\tau(M)$ and the τ–small submodules of M.

Proposition 2.2.4 $J_\tau(M)$ *contains every* τ*–small submodule of* M.

Proof. If $J_\tau(M) = M$, there is nothing to prove, so suppose X is a τ–maximal submodule of M and let N be a τ–small submodule of M. If $X \subset N^c + X$, then, by Proposition 2.2.2, $N^c + X \in F_\tau(M)$ which is a contradiction. Consequently, $X = N^c + X$ in which case $N \subseteq N^c \subseteq X$. □

Proposition 2.2.5 *Let* M *be an* R*–module, then:*

1. $M/J_\tau(M)$ *is a τ–radical free* R*–module.*
2. *The following conditions are equivalent:*
   (a) *M is a τ–radical free* R*–module.*
   (b) *If* $m \in M$, $m \neq 0$, *there is a τ–simple* R*–module* N *and an* R*–homomorphism* $f : M \to N$ *such that* $f(m) \neq 0$.
   (c) *The map* $\varphi : M \to \prod_{\alpha \in \Delta} M/N_\alpha : m \to (m + N_\alpha)$, *where* $\{ N_\alpha \}_{\alpha \in \Delta}$ *is the non–empty family of τ–maximal submodules of* M, *is a monomorphism.*
   (d) M *can be embedded in a product of τ–simple* R*–modules.*

Proof. 1. Since a submodule N of M is τ–maximal in M if and only if $N/J_\tau(M)$ is τ–maximal in $M/J_\tau(M)$, the family $\{ N_\alpha \}_{\alpha \in \Delta}$ contains all the τ–maximal submodules of M if and only if the family $\{ N_\alpha/J_\tau(M) \}_{\alpha \in \Delta}$ contains all the τ–maximal submodules of $M/J_\tau(M)$. Thus, $J_\tau(M/J_\tau(M)) = \cap_{\alpha \in \Delta} (N_\alpha /J_\tau(M)) = \left(\cap_{\alpha \in \Delta} N_\alpha \right) /J_\tau(M) = J_\tau(M)/J_\tau(M) = 0$.

2. (a) ⇒ (b). Let $m \in M$, $m \neq 0$. Since $J_\tau(M) = 0$, there is a τ–maximal submodule N of M such that $m \notin N$. But then M/N is τ–simple and the R–homomorphism $f : M \to M/N : x \to x + N$ is such that $f(m) \neq 0$.

(b) ⇒ (a). Suppose m is any nonzero element of M and let the R–homomorphism $f : M \to N$ be such that $f(m) \neq 0$ where N is a τ–simple R–module. Since nonzero submodules of τ–simple modules are τ–simple, ker f is a τ–maximal submodule of M. Thus, $J_\tau(M) \subseteq \ker f$ and so

$m \notin J_\tau(M)$. Therefore, $J_\tau(M) = 0$.

(b) $\Rightarrow$ (c). If $m \in M$ and $\varphi(m) = 0$, then $(m + N_\alpha) = 0$. Thus, m is in every τ–simple submodule of M and so $m \in J_\tau(M) = 0$. Hence, φ is a monomorphism as was asserted.

(c) $\Rightarrow$ (d) is obvious.

(d) $\Rightarrow$ (b). Let $\varphi : M \to \prod_{\alpha \in \Delta} N_\alpha$, where $\{N_\alpha\}_{\alpha \in \Delta}$ is a family of τ–simple R–modules, be an R–monomorphism. If $m \in M$, $m \neq 0$, then $\varphi(m) \neq 0$ and so $\pi_\beta \circ \varphi(m) \neq 0$ for some $\beta \in \Delta$ where $\pi_\beta : \prod_{\alpha \in \Delta} N_\alpha \to N_\beta$ is the canonical projection. $\pi_\beta \circ \varphi$ is the required homomorphism. □

It follows from Definition 2.2.1 that $J_\tau(M) = \cap_{g \in \Delta} \ker g$ where $\Delta = \{g \mid g : M \to X$ is an R–epimorphism and X is a τ–simple R–module $\}$. We adopt the usual convention that if $\Delta = \emptyset$, then $\cap_{g \in \Delta} \ker g = M$.

Proposition 2.2.6 *The following hold from any* R*–modules* M *and* N.

1. *If* $f : M \to N$ *is an* R*–homomorphism,* $f(J_\tau(M)) \subseteq J_\tau(N)$.
2. $J_\tau(R)$ *is an ideal of* R.
3. $MJ_\tau(R) \subseteq J_\tau(M)$.

Proof. 1. Let $\Delta = \{g \mid g : N \to X$ is an R–epimorphism and X is a τ–simple R–module $\}$. If $g \in \Delta$ and $g \circ f = 0$, then clearly $f(J_\tau(M) \subseteq \ker g$. If $g \circ f \neq 0$, then $g \circ f(M)$ is a nonzero submodule of the τ–simple module X and so is τ–simple. Since $g \circ f : M \to g \circ f(M)$ is an R–epimorphism, it follows that $J_\tau(M) \subseteq \ker g \circ f$. Consequently, $f(J_\tau(M)) \subseteq \ker g$ and so it must be that case that $f(J_\tau(M)) \subseteq J_\tau(N)$.

2. If $r \in R$, then by 1, $f_r(J_\tau(R)) \subseteq J_\tau(R)$ where $f_r : R \to R : x \to rx$. Hence, $r\,J_\tau(R) \subseteq J_\tau(R)$ for each $r \in R$.

3. If X is a τ–simple R–module and $x \in X$, define $f : R \to X$ by $f(r) = xr$.

Then by 1, $f(J_\tau(R)) \subseteq J_\tau(X) = 0$ and so it follows that $XJ_\tau(R) = 0$. Next, if $g : M \to X$ is an R–epimorphism in Δ, $g(MJ_\tau(R)) = XJ_\tau(R) = 0$ and so $MJ_\tau(R) \subseteq \ker g$. Therefore, $MJ_\tau(R) \subseteq J_\tau(M)$. □

Corollary 2.2.7 $MJ_\tau(R) = 0$ *for every* τ*–radical free* R*–module* M.

Since a τ–simple R–module is τ-radical free, Corollary 2.2.7 shows that $MJ_\tau(R) = 0$ for every τ–simple R–module M. Thus, $J_\tau(R) \subseteq \cap_{\alpha \in \Delta}(0 : N_\alpha)$ where $\{N_\alpha\}_{\alpha \in \Delta}$ is the family of all τ–simple R–modules. Conversely, if $r \in \cap_{\alpha \in \Delta}(0 : N_\alpha)$, then $(R/K)r = 0$ where K is any τ–maximal right ideal of R. But since R has an identity, $r \in K$. Hence, $r \in K$ for every τ–maximal right ideal K of R and so $r \in J_\tau(R)$. Thus, we have the following proposition.

Proposition 2.2.8 $J_\tau(R) = \cap_{\alpha \in \Delta}(0 : N_\alpha)$ *where* $\{N_\alpha\}_{\alpha \in \Delta}$ *is the family of all* τ*–simple* R*–modules.*

Proposition 2.2.5 shows that $R/J_\tau(R)$ is a τ–radical free R–module. τ can be used to induce a torsion theory σ on Mod–$R/J_\tau(R)$ so that $R/J_\tau(R)$ will be will be a σ–radical free as an $R/J_\tau(R)$–module.

Proposition 2.2.9 *Let* $\eta : R \to S$ *be a ring epimorphism and suppose that* F(R) *is a Gabriel filter of right ideals of* R. *Then* $\eta(F(R)) = \{$ K is a right ideal of S $\mid \eta^{-1}(K) \in F(R)\}$ *is a Gabriel filter of right ideals of* S.

Proof. If A and B are non–empty subsets of a ring R, $(A : B)_R$ will denote the set of all elements $r \in R$ such that $Br \subseteq A$. To complete the proof, we need to show that the two conditions of Definition 1.1.12 are satisfied. For condition 1, suppose $K \in \eta(F(R))$, $y \in S$ and let $x \in R$ be such that $\eta(x) = y$. Then $(\eta^{-1}(K) : x)_R \in F(R)$ and $(\eta^{-1}(K) : x)_R = \eta^{-1}((K : y)_S)$. Hence, $(K : y)_S \in \eta(F(R))$. For condition 2, let $J \in \eta(F(R))$ and suppose that K is a right ideal of S such that $(K : y)_S \in \eta(F(R))$ for all $y \in J$. We need to show that $K \in \eta(F(R))$. Since $(K : y)_S \in \eta(F(R))$, $\eta^{-1}((K : y)_S) \in F(R)$. But $(\eta^{-1}(K) : x)_R = \eta^{-1}((K : y)_S)$ where $x \in \eta^{-1}(J)$ is such that $\eta(x) = y$. Hence for any $x \in \eta^{-1}(J) \in F(R)$, $(\eta^{-1}(K) : x)_R \in F(R)$ and so $\eta^{-1}(K) \in F(R)$ which shows that $K \in \eta(F(R))$. □

If $\eta : R \to S$ is a ring epimorphism and τ is a torsion theory on Mod–R, the torsion theory on Mod–S induced by τ will be denoted by $\eta(\tau)$. Note that if I is an ideal of R, then it follows from Proposition 2.2.9 that the set ideals $F(R/I) = \{ K/I \mid K \supseteq I \text{ and } K \in F(R) \}$ is a Gabriel filter of right ideals of R/I.

Proposition 2.2.10 *Let* τ *is a torsion theory on* Mod–R. *Then:*

1. *If* I *is any ideal of* R *such that* $I \subseteq J_\tau(R)$, *then* $J_{\pi(\tau)}(R/I) = J_\tau(R)/I$ *where* $\pi : R \to R/I$ *is the canonical ring homomorphism.*
2. $J_{\pi(\tau)}(R/t_\tau(R)) = J_\tau(R)/t_\tau(R)$ *where* $t_\tau(R)$ *is the* τ*–torsion ideal of* R *and* $\pi : R \to R/t_\tau(R)$ *is the canonical ring homomorphism.*
3. $R/J_\tau(R)$ *is a* $\pi(\tau)$*–radical free* $R/J_\tau(R)$*–module where* $\pi : R \to R/J_\tau(R)$ *is the canonical ring homomorphism.*

Proof. 1, 2, and 3 follow directly from the observation that if K is an ideal of R and $\pi : R \to R/K$ is the canonical ring homomorphism, the $\pi(\tau)$–maximal right ideals of ring R/K is are precisely those τ–maximal right ideals of R which contain K. □

Proposition 2.2.11 *If* $\{ M_\alpha \}_{\alpha \in \Delta}$ *is any family of* R–*modules, then* $J_\tau(\oplus_{\alpha \in \Delta} M_\alpha) = \oplus_{\alpha \in \Delta} J_\tau(M_\alpha)$.

Proof. We prove the case for $\Delta = \{ 1, 2 \}$. The argument for the general case is similar. Since $M_1 \subseteq M_1 \oplus M_2$ and $M_2 \subseteq M_1 \oplus M_2$, 1 of Proposition 2.2.6 shows that $J_\tau(M_1) \subseteq J_\tau(M_1 \oplus M_2)$ and $J_\tau(M_2) \subseteq J_\tau(M_1 \oplus M_2)$. Thus, $J_\tau(M_1) \oplus J_\tau(M_2) \subseteq J_\tau(M_1 \oplus M_2)$. Since $(M_1 \oplus M_2)/(N_1 \oplus M_2) \to M_1/N_1 :$ $(m_1, m_2) + N_1 \oplus M_2 \to m_1 + N_1$ is an isomorphism, if N_1 is a τ–maximal submodule of M_1, $N_1 \oplus M_2$ is a τ–maximal submodule of $M_1 \oplus M_2$. Hence, $(m_1, m_2) \in J_\tau(M_1 \oplus M_2)$ leads to $(m_1, m_2) \in N_1 \oplus M_2$ and so $m_1 \in N_1$. Thus, $m_1 \in J_\tau(M_1)$. Similarly, $m_2 \in J_\tau(M_2)$ and so equality holds. □

Corollary 2.2.12 *A direct sum of* R–*modules is* τ–*radical free if and only if each summand is* τ–*radical free.*

2.3 Noetherian Modules Relative to a Torsion Theory

Definition 2.3.1 A set $\mathcal{S}$ of submodules of an R–module M satisfies the *ascending chain condition* if for each ascending chain $M_1 \subseteq M_2 \subseteq ... \subseteq M_k \subseteq ...$ of submodules of M each of which is in $\mathcal{S}$, there is a positive integer n such that $M_k = M_{k+1}$ for all $k \geq n$. For a torsion theory τ on Mod–R, an R–module M is said to be τ–*noetherian* if the set $\mathcal{P}_\tau(M)$ of τ–pure submodules of M satisfies the ascending chain condition. R is (right) τ–noetherian if R is τ–noetherian as an R–module.

Definition 2.3.2 An R–module M is said to be τ*–finitely generated* if M possesses a finitely generated τ–dense submodule.

Proposition 2.3.3 *The following are equivalent for an* R*–module* M.

1. M *is* τ*–noetherian.*
2. *If* $M_1 \subseteq M_2 \subseteq \cdots \subseteq M_k \subseteq \cdots$ *is an ascending chain of submodules of* M, *then there is a positive integer* n *such that* $M_{k+1}/M_k \in \mathscr{T}$ *for all* $k \geq n$.
3. *If* $\mathscr{S}$ *is a non–empty set of submodules of* M, *there is an* $N \in \mathscr{S}$ *which is* τ*–dense in every* $X \in \mathscr{S}$ *such that* $N \subseteq X$.
4. *If* N *is a submodule of* M, *then every non–empty collection of modules in* $\mathscr{P}_\tau(N)$ *has a maximal element.*
5. *Every submodule of* M *is* τ*–finitely generated.*
6. *The set of* τ*–pure,* τ*–finitely generated submodules of* M *satisfies the ascending chain condition.*

Proof. $1 \Rightarrow 2$. Suppose $M_1 \subseteq M_2 \subseteq \cdots \subseteq M_k \subseteq \cdots$ is an ascending chain of submodules of M. Then $M_1^c \subseteq M_2^c \subseteq \cdots \subseteq M_k^c \subseteq \cdots$ is an ascending chain of τ–pure submodules of M and so there exists a positive integer n such that $M_k^c = M_{k+1}^c$ for all $k \geq n$. But for any k, the intersection of the τ–pure submodules of M which contain M_k is precisely $t_\tau(M/M_k) = M_k^c/M_k$. Thus, $M_k^c = M_{k+1}^c$ together with $t_\tau(M/M_k) = M_k^c/M_k = M_{k+1}^c/M_k$ implies that $M_{k+1}/M_k \subseteq M_{k+1}^c/M_k$ is τ–torsion for all $k \geq n$.

$2 \Rightarrow 3$. Assume 3 is false. Then a non–empty set of submodules of M exists for which the condition given in 3 does not hold. Let $\mathscr{S}$ be such a set. Then for any $N \in \mathscr{S}$ there is an $X \in \mathscr{S}$ such that $N \subset X$ and $N \notin F_\tau(X)$. Thus, if $M_1 \in \mathscr{S}$, there must exist an element M_2 of $\mathscr{S}$ such that $M_1 \subset M_2$

and $M_1 \notin F_\tau(M_2)$. Similarly, there is an $M_3 \in \mathscr{S}$ such that $M_2 \subset M_3$ and $M_2 \notin F_\tau(M_3)$. Inductively, we obtain a strictly increasing chain $M_1 \subset M_2 \subset M_3 \subset \ldots \subset M_k \subset \ldots$ of submodules of M such that $M_k \notin F_\tau(M_{k+1})$ for any $k \geq 1$. Therefore, 2 is false and so 2 implies 3.

$3 \Rightarrow 4$. Let N be a submodule of M and suppose $\mathscr{A}$ is a non–empty collection submodules of N each of which is τ–pure in N. If $\mathscr{A}$ does not have a maximal element, then we can build a strictly increasing chain $N_1 \subset N_2 \subset \cdots$ of submodules from $\mathscr{A}$. If $\mathscr{S}$ is the set of all modules in this chain, then, by 3, there is a positive integer n such that N_n is τ–dense in N_k for all $k \geq n$. Thus, $N_k / N_n \in \mathscr{T}$ for all $k \geq n$. But N_n is τ–pure in N and so $N_k / N_n \in \mathscr{F}$ Hence, $N_k / N_n \in \mathscr{T} \cap \mathscr{F} = 0$ and so $N_k = N_n$ for all $n \geq k$. But this contradicts the fact that the chain is strictly increasing. Consequently, $\mathscr{A}$ must have a maximal element.

$4 \Rightarrow 5$. Let N be any submodule of M. We need to show that N is τ–finitely generated. If N is finitely generated, N is τ–finitely generated since $N \in F_\tau(N)$. If N is not finitely generated, let $\mathscr{S}$ be the set of all finitely generated submodules of N. Let $\mathscr{S}^c = \{ X^c \mid X \in \mathscr{S} \}$ where the τ–pure closure is taken with respect to N. Then by 4, $\mathscr{S}^c$ has a maximal element, say $(x_1R + \cdots x_nR)^c$. If $(x_1R + \cdots + x_nR)^c = N$, N will be τ–finitely generated since $x_1R + \cdots + x_nR$ is τ–dense in $(x_1R + \cdots + x_nR)^c$. Suppose $(x_1R + \cdots + x_nR)^c \neq N$ and let $x \in N \setminus (x_1R + \cdots + x_nR)^c$. Then $(x_1R + \cdots x_nR)^c$ is a proper submodule of $(x_1R + \cdots + x_nR + xR)^c$ in which case $(x_1R + \cdots + x_nR + xR)^c = N$ by the maximality of $(x_1R + \cdots + x_nR)^c$. But $x_1R + \cdots + x_nR + xR$ is τ–dense in $(x_1R + \cdots + x_nR + xR)^c$ and so N is τ–finitely generated.

$5 \Rightarrow 6$. Let $M_1 \subseteq M_2 \subseteq \cdots \subseteq M_k \subseteq \cdots$ be an increasing chain of τ–pure, τ–finitely generated submodules of M. Consider the τ–finitely generated submodule $\cup_{k=1}^{\infty} M_k$ of M and by 5 let X be a finitely generated τ–dense submodule of $\cup_{k=1}^{\infty} M_k$. Since X is finitely generated, there is a positive integer n such that $X \subseteq M_k$ for all $k \geq n$. Moreover, X is τ–dense in each such M_k. Consequently, $M_{k+1}/M_k \cong (M_{k+1}/X)/(M_k/X) \in \mathcal{T}$. But M_k is τ–pure in M and so $M_{k+1}/M_k \in \mathcal{F}$. Hence, $M_k = M_{k+1}$ for all $k \geq n$.

$6 \Rightarrow 1$. Let $M_1 \subseteq M_2 \subseteq \cdots \subseteq M_k \subseteq \cdots$ be an ascending chain of sub–modules from $\mathcal{P}_\tau(M)$. Choose $x_1 \in M_1$ and for $k > 1$ choose $x_k \in M_k$ as follows: if $M_k \neq M_{k-1}$, let $x_k \in M_k \setminus M_{k-1}$, otherwise choose $x_k = 0$. Now consider the chain $(x_1R)^c \subseteq (x_1R + x_2R)^c \subseteq \cdots$ of τ–pure, τ–finitely generated submodules of M. By assumption there is a positive integer n such that $(x_1R + \cdots + x_kR)^c = (x_1R + \cdots + x_nR)^c$ for all $k \geq n$. Notice next that $(x_1R + \cdots + x_kR)^c \subseteq M_k$ for each $k \geq 1$ since M_k is a τ–pure submodule of M. Consequently, $(x_1R + \cdots + x_kR)^c = (x_1R + \cdots + x_nR)^c$ implies that $x_k \in M_n$ for each $k \geq n$. But this impossible in view of how the x_k were chosen unless $x_k = 0$ and $M_k = M_n$ for all $k \geq n$. □

Corollary 2.3.4 *Every τ–noetherian R–module has a τ–maximal submodule.*

Proposition 2.3.5 *If N is a submodule of an R–module M, then M is τ–noetherian if and only if N and M/N are τ–noetherian.*

Proof. First, assume that M is τ–noetherian. If $X/N \in \mathcal{P}_\tau(M/N)$, then $X \in \mathcal{P}_\tau(M)$. Consequently, any ascending chain of modules in $\mathcal{P}_\tau(M/N)$ gives rise to an ascending chain of modules in $\mathcal{P}_\tau(M)$ which will stabilize. Thus, the original chain will stabilize and so M/N is τ–noetherian. Next, consider the mapping $\mathcal{P}_\tau(N) \to \mathcal{P}_\tau(M) : X \to X^c$ where the τ–pure closure of

X is taken with respect to M. We claim this mapping is injective. First, note if $X \in \mathscr{P}_\tau(N)$, then $X \subseteq N \cap X^c \subseteq N$ so that $(N \cap X^c)/X \subseteq N/X \in \mathscr{F}$. Likewise, $X \subseteq N \cap X^c \subseteq X^c$ so that $(N \cap X^c)/X \subseteq X^c/X = t_\tau(M/X) \in \mathscr{T}$. Therefore, $X = N \cap X^c$ and so if $X, Y \in \mathscr{P}_\tau(N)$ and $X^c = Y^c$, then $X = N \cap X^c = N \cap Y^c = Y$. Hence, the map is injective. This is sufficient to show that N is τ–noetherian.

Conversely, suppose that both N and M/N are τ–noetherian and let $M_1 \subseteq M_2 \subseteq \cdots \subseteq M_k \subseteq \cdots$ be an ascending chain in $\mathscr{P}_\tau(M)$. First, note that $N/(N \cap M_k) \cong (N + M_k)/M_k \in \mathscr{F}$ and so $N \cap M_k$ is τ–pure in N. Hence, $N \cap M_1 \subseteq N \cap M_2 \subseteq \cdots \subseteq N \cap M_k \subseteq \cdots$ is an ascending chain in $\mathscr{P}_\tau(N)$ and so there is a positive integer m such that $N \cap M_k = N \cap M_m$ for all $k \geq m$. If $(N + M_k)^c$ is the τ–pure closure of $N + M_k$ in M for $k = 1, 2, \cdots$, then $(N + M_1)^c/N \subseteq (N + M_2)^c/N \subseteq \cdots \subseteq (N + M_k)^c/N \subseteq \cdots$ is an ascending chain in $\mathscr{P}_\tau(M/N)$. But M/N is τ–noetherian and so there is a positive integer n such that $(N + M_k)^c/N = (N + M_n)^c/N$ for all $k \geq n$. It now follows that $(N + M_k)/(N + M_n)$ is isomorphic to a submodule of $(N + M_n)^c/(N + M_n) \in \mathscr{T}$ and so $(N + M_k)/(N + M_n) \in \mathscr{T}$ whenever $k \geq n$. If $k_0 = \max(m, n)$, then $(N + M_k)/(N + M_{k_0}) \cong ((N + M_k)/N)/((N + M_{k_0})/N) \cong$ $(M_k/(N \cap M_k))/(M_{k_0}/(N \cap M_{k_0})) \cong (M_k/(N \cap M_{k_0}))/(M_{k_0}/(N \cap M_{k_0})) \cong$ $M_k/M_{k_0} \in \mathscr{F}$. Hence, $M_k/M_{k_0} \in \mathscr{T} \cap \mathscr{F} = 0$ and so it must be the case that $M_k = M_{k_0}$ when $k \geq k_0$. □

Corollary 2.3.6 *If* $0 \to L \to M \to N \to 0$ *is a short exact sequence of* R–*modules, then* M *is* τ–*noetherian if and only if* L *and* N *are* τ–*noetherian.*

If $M \in \mathscr{T}$ or if M is τ–simple, $\mathscr{P}_\tau(M) = \{ M \}$ or $\mathscr{P}_\tau(M) = \{ 0, M \}$, respectively, and so every ascending chain in $\mathscr{P}_\tau(M)$ clearly stabilizes. Thus, M is τ–noetherian. The following corollary to Proposition 2.3.5 now follows immediately.

Corollary 2.3.7 *The following are equivalent for an* R–*module* M.

1. M *is* τ–*noetherian.*
2. M/N *is* τ–*noetherian for every* $N \in \mathscr{T}$.
3. M/t(M) *is* τ–*noetherian.*

Corollary 2.3.8 *Any finite direct product* $\prod_{k=1}^{n} M_k$ *of* R–*modules is* τ–*noetherian if and only if* M_k *is* τ–*noetherian for* $k = 1, 2, \cdots, n$.

Proof. The proof is by induction on n. □

Proposition 2.3.9 *If* N *is a submodule of an* R–*module* M, *then* N *is* τ–*noetherian if and only if* N^c *is* τ–*noetherian.*

Proof. Suppose that N is τ–noetherian. Since $N^c/N \in \mathscr{T}$, N^c/N is clearly τ–noetherian. Hence, it follows from Proposition 2.3.5 that N^c is τ–noetherian. The converse also follows directly from Proposition 2.3.5. □

Proposition 2.3.10 R *is* τ–*noetherian if and only if every* τ–*finitely generated* R–*module is* τ–*noetherian.*

Proof. Suppose R is τ–noetherian and let M be any τ–finitely generated R–module. If N is a finitely generated τ–dense submodule of M, there is a positive integer n and an R–epimorphism $R^n \to N$. Let $X \subseteq R^n$ be such that $R^n/X \cong N$. Corollary 2.3.8 shows that R^n is τ–noetherian and so R^n/X is τ–noetherian by Proposition 2.3.5. But $N^c = M$ and so M is τ–noetherian by Proposition 2.3.9. The converse follows since R has an identity. □

We conclude this section with a relative form of Nakayama's Lemma due to Porter [32]. The proof of Nakayama's Lemma reveals some of the difficulties in formulating a relative version of this result.

Proposition 2.3.11 (Nakayama's Lemma) *For any right ideal* K *of* R, *the following are equivalent:*

1. $K \subseteq J(R)$. ($J(R)$ = *the Jacobson radical of* R.)
2. *For any finitely generated* R–*module* M, $MK = M$ *implies* $M = 0$.
3. *For any submodule* N *of* M *such that* M/N *is finitely generated,* $N + MK = M$ *implies* $N = M$.

Proof. $1 \Rightarrow 2$. If $M \neq 0$, partial order the set $\mathscr{S}$ of proper submodules of M by inclusion. If $M_1 \subseteq M_2 \subseteq \cdots \subseteq M_k \subseteq \cdots$ is a chain in $\mathscr{S}$, $\cup_{k=1}^{\infty} M_k$ is a proper submodule of M due to the fact that M is finitely generated. Thus, by Zorn's Lemma, $\mathscr{S}$ has at least one maximal element. If M' is a maximal element of $\mathscr{S}$, M/M' is simple and so $(M/M')K = 0$. Thus, $MK \subseteq M'$ which tells us that $MK \neq M$. Hence, $MK = M$ implies $M = 0$.

$2 \Rightarrow 3$. It suffices to show $(M/N)K = M/N$ since M/N is finitely generated. If $m + N \in M/N$, then because $M = N + MK$ there are elements $n \in N$ and $\sum_{i=1}^{n} m_i k_i \in MK$ such that $m + N = n + \left(\sum_{i=1}^{n} m_i k_i \right) + N = \left(\sum_{i=1}^{n} m_i k_i \right) + N = \sum_{i=1}^{n} (m_i + N) k_i \in (M/N)K$.

$3 \Rightarrow 1$. If $K \not\subseteq J(R)$, there must exist a maximal right ideal M of R such that $M + K = R$. Thus, $M + RK = R$. But R/M is finitely generated and so $M = R$, a contradiction. Hence, $K \subseteq J(R)$. □

In the proof of $1 \Rightarrow 2$ of Nakayama's Lemma we have seen that a finitely generated R–module has at least one maximal submodule. One difficulty in extending Nakayama's Lemma to a torsion theoretical setting is that τ–finitely generated R–modules may fail to have τ–maximal submodules. Following Porter, we establish the following definition:

Definition 2.3.12 The torsion theory τ is said to satisfy τ–*Max* if every τ–finitely generated R–module such that $\mathscr{P}_\tau(M) \neq \{ M \}$ has τ–maximal submodules. $Max_\tau(M)$ and $Max_\tau(R)$ will denote the set of τ–maximal submodules of M and the set of τ–maximal right ideals of R, respectively.

As has been noted in Proposition 2.1.2, if there is an R–module M such that $Max_\tau(M) \neq \emptyset$, then $Max_\tau(R) \neq \emptyset$. Corollary 2.3.4 and Proposition 2.3.10 show that if R is τ–noetherian, then $Max_\tau(M) \neq \emptyset$ for every τ–finitely generated R–module M.

Proposition 2.3.13 *If* M *is a τ–finitely generated* R*–module, then* M/N *is τ–finitely generated for any submodule* N *of* M.

Proof. If X is a finitely generated and τ–dense submodule of M, then

$(N + X)/N$ is a finitely generated and τ–dense submodule of M/N. □

The following proposition is a generalized version of a proposition which appears in [4].

Proposition 2.3.14 *If* τ *satisfies* τ–Max, *the following are equivalent for any submodule* X *of a* τ*–finitely generated* R*–module* M.

1. $X \subseteq J_\tau(M)$.
2. *If* $H \subseteq M$, *then* $M/(X + H) \in \mathscr{T}$ *implies that* $M/H \in \mathscr{T}$.

Proof. 1 ⇒ 2. If $H = M$ there is nothing to prove, so suppose $H \subset M$, $M/(X + H) \in \mathscr{T}$ and $M/H \notin \mathscr{T}$. By Proposition 2.3.13, M/H is τ–finitely generated and we know that $\mathscr{P}_\tau(M/H) \neq \{ M/H \}$ since $t_\tau(M/H) \neq M/H$. Consequently, since τ–Max holds, $\text{Max}_\tau(M/H) \neq \emptyset$. If $N/H \in \text{Max}_\tau(M/H)$, M/N is τ–simple and so $X \subseteq N$ which gives $X + H \subseteq N$. But M/N is a homomorphic image of $M/(X + H) \in \mathscr{T}$ and so $M/N \in \mathscr{T}$. Now this cannot be the case since $M/N \neq 0$ and $M/N \in \mathscr{F}$. Consequently, when $M/(X + H) \in \mathscr{T}$, $M/H \in \mathscr{T}$.

2 ⇒ 1. Let $N \in \text{Max}_\tau(M)$. Then by Proposition 2.2.2, N is maximal among the submodules of M which are not in $F_\tau(M)$. If $X \not\subseteq N$, $X + N \in F_\tau(M)$ and so $M/(X + N) \in \mathscr{T}$. By assumption this implies that $M/N \in \mathscr{T}$ and so it follows that $N = M$, a contradiction. Hence, $X \subseteq N$. □

Now for a relative form of Nakayama's Lemma.

Proposition 2.3.15 (The Generalized Nakayama Lemma) *Suppose* τ *satisfies* τ–Max. *For any right ideal* K *of* R, *the following are equivalent:*

1. $K \subseteq J_\tau(R)$.
2. *For any* τ*–finitely generated* R*–module* M, $M/MK \in \mathscr{T}$ *implies that* $M \in \mathscr{T}$.
3. *If* M *is* τ*–finitely generated and* $N \subseteq M$, *then* $M/(N + MK) \in \mathscr{T}$ *implies that* $M/N \in \mathscr{T}$.

Proof. $1 \Rightarrow 3$. By Proposition 2.2.6, $MJ_\tau(R) \subseteq J_\tau(M)$ and so $MK \subseteq J_\tau(M)$ when $K \subseteq J_\tau(R)$. Therefore, the implication follows from Proposition 2.3.14.

$3 \Rightarrow 1$. Since R has an identity, R is τ–finitely generated. Hence, $Max_\tau(R) \neq \emptyset$ and so $J_\tau(R) \neq R$. If $K \not\subseteq J_\tau(R)$, there is an $X \in Max_\tau(R)$ such that $K \not\subseteq X$. But by Proposition 2.2.2, X is maximal among those right ideals of R such that $X \notin F_\tau(R)$. Hence, $X + K \in F_\tau(R)$ and so $X + RK \in F_\tau(R)$ which implies that $R/(X + RK) \in \mathscr{T}$. Now $R/X \in \mathscr{F}$ and so it cannot be the case that $R/X \in \mathscr{T}$ since $X \neq R$. Thus, when $K \not\subseteq J_\tau(R)$, 3 fails to hold and so $3 \Rightarrow 1$.

$2 \Rightarrow 3$. If $M/(N + MK) \in \mathscr{T}$, it follows that $(M/N)/[MK/(N \cap MK)]$ is in $\mathscr{T}$. Moreover, $MK/(N \cap MK)$ embeds in $(M/N)K$ via the map $\Sigma_{i=1}^{q} m_i k_i + (N \cap MK) \to \Sigma_{i=1}^{q} (m_i + N)k_i$. Hence, $(M/N)/[(M/N)K]$ is a homomorphic image of $(M/N)/[MK/(N \cap MK)]$ and so $(M/N)/[(M/N)K]$ is in $\mathscr{T}$. Consequently by 2, $M/N \in \mathscr{T}$.

$3 \Rightarrow 2$. Let $N = 0$. □

Corollary 2.3.16 *Let* τ *satisfy* τ–Max *and suppose* M *is a* τ*–finitely generated* R*–module and that* K *is an ideal of* R *such that* $K \subseteq J_\tau(R)$. If $M \otimes_R (R/K)$ *is in* $\mathscr{T}$, *then* $M \in \mathscr{T}$.

Proof. $M \otimes_R (R/K) \cong M/MK$ and so the result follows from the proposition.

□

2.4. Artinian Modules Relative to a Torsion Theory

Definition 2.4.1 A set $\mathscr{S}$ of submodules of an R–module M is said to satisfy the *descending chain condition* if for each descending chain $M_1 \supseteq M_2 \supseteq ... \supseteq M_k \supseteq ...$ of submodules of M each of which is in $\mathscr{S}$, there is a positive integer n such that $M_k = M_{k+1}$ for all $k \geq n$. For a torsion theory τ on Mod–R, an R–module M is said to be τ–*artinian* if the set $\mathscr{P}_\tau(M)$ of τ–pure submodules of M satisfies the descending chain condition. A ring R is (right) τ–artinian if R is τ–artinian as an R–module.

If $M \in \mathscr{T}$ or M is τ–simple, then $\mathscr{P}_\tau(M) = \{ M \}$ or $\mathscr{P}_\tau(M) = \{ 0, M \}$, respectively. Thus, M is clearly τ–artinian.

Definition 2.4.2 An R–module M is said to be τ–*cofinitely generated* if, whenever $\{ M_\beta \mid \beta \in \mathscr{B} \}$ is a set of τ–pure submodules of M such that $\cap_{\beta \in \mathscr{B}} M_\beta = t_\tau(M)$, there is a finite subset $\mathscr{A} \subseteq \mathscr{B}$ such that $\cap_{\alpha \in \mathscr{A}} M_\beta = t_\tau(M)$.

Proposition 2.4.3 *For any* R–*module* M, *the following are equivalent:*

1. M *is* τ–*artinian.*
2. *For any descending chain* $M_1 \supseteq M_2 \supseteq \cdots \supseteq M_k \supseteq \cdots$ *of submodules of* M, *there is a positive integer* n *such that* $M_k / M_{k+1} \in \mathscr{T}$ *for all* $k \geq n$.
3. *If* $\mathscr{S}$ *is a non–empty set of submodules of* M, *then there exists an* $N \in \mathscr{S}$ *such that if* $X \in \mathscr{S}$ *and* $X \subseteq N$, *then* X *is* τ–*dense in* N.

4. *If* N *is a submodule of* M, *then any collection of modules in* $\mathscr{P}_\tau(N)$ *has a minimal element.*
5. *Modules in* $\mathscr{P}_\tau(M)$ *satisfy the finite intersection property.*
6. *Every* (τ*–torsionfree*) *homomorphic image of* M *is* τ*–cofinitely generated.*

Proof. $1 \Rightarrow 2$. If $M_1 \supseteq M_2 \supseteq \cdots \supseteq M_k \supseteq \cdots$ is a decreasing chain of submodules of M, then $M_1^c \supseteq M_2^c \supseteq \cdots \supseteq M_k^c \supseteq \cdots$ is a decreasing chain of modules in $\mathscr{P}_\tau(M)$. Hence, there is a positive integer n such that $M_k^c = M_{k+1}^c$ for all $k \geq n$. But for any k the intersection of all the τ–pure submodules of M which contain M_k is exactly $t_\tau(M/M_k)$. Hence, $t_\tau(M/M_k) = M_k^c/M_k = M_{k+1}^c/M_k$ implies that $M_{k+1}/M_k \in \mathscr{T}$.

$2 \Rightarrow 3$. Assume that 3 is false. Let $\mathscr{S}$ be a non–empty set of submodules of M for which the condition given in 3 does not hold. Then for any $N \in \mathscr{S}$ there is an $X \in \mathscr{S}$ such that $X \subseteq N$ and $X \notin F_\tau(N)$. Thus, if $M_1 \in \mathscr{S}$, there must exist an $M_2 \in \mathscr{S}$ such that $M_1 \supseteq M_2$ and $M_2 \notin F_\tau(M_1)$. Similarly, there is an $M_3 \in \mathscr{S}$ such that $M_2 \supseteq M_3$ and $M_3 \notin F_\tau(M_2)$. Continuing in this fashion we obtain a decreasing chain of submodules of M for which $M_{k+1} \notin F_\tau(M_k)$ for any $k \geq 1$. Therefore, 2 is false and so 2 implies 3.

$3 \Rightarrow 4$. Let N be as submodule of M and suppose that $\mathscr{A}$ is a non–empty collection of τ–pure submodules of N. If $\mathscr{A}$ does not have a minimal element, we can find a strictly decreasing chain $N_1 \supset N_2 \supset \cdots \supset N_k \supset \cdots$ of submodules from $\mathscr{A}$. If $\mathscr{S}$ is the set of all submodules of N in this chain, then, by 3, there is a positive integer n such that $N_k \in F_\tau(N_n)$ for all $k \geq n$. Hence, $N_n/N_k \in \mathscr{T}$ for all $k \geq n$. But $N_k \in \mathscr{P}_\tau(N)$ and so

$N_n/N_k \in \mathscr{F}$ for all $k \geq n$. Thus, $N_n/N_k \in \mathscr{T} \cap \mathscr{F} = 0$ and so $N_k = N_n$ for all $k \geq n$. But this contradicts the fact that $N_n \supset N_k$ for all $k \geq n$.

$4 \Rightarrow 5$. Let $\{ M_\beta \}_{\beta \in \mathscr{B}}$ be a family of submodules of M each of which is in $\mathscr{P}_\tau(M)$. Consider $\cap_{\beta \in \mathscr{B}} M_\beta$. Let $\mathscr{S}$ be the collection of all finite intersections of modules in the family $\{ M_\beta \}_{\beta \in \mathscr{B}}$. Then by 4, $\mathscr{S}$ has a minimal element, say $\cap_{\alpha \in \mathscr{A}} M_\alpha$ where $\mathscr{A}$ is a finite subset of $\mathscr{B}$. If $\cap_{\beta \in \mathscr{B}} M_\beta = \cap_{\alpha \in \mathscr{A}} M_\alpha$, we are finished, so suppose $\cap_{\beta \in \mathscr{B}} M_\beta \neq \cap_{\alpha \in \mathscr{A}} M_\alpha$. If $m \in \cap_{\alpha \in \mathscr{A}} M_\alpha \setminus \cap_{\beta \in \mathscr{B}} M_\beta$, there is a $\gamma \in \mathscr{B}$ such that $m \notin M_\gamma$. Then $\cap_{\alpha \in \mathscr{A}} M_\alpha \supset \left(\cap_{\alpha \in \mathscr{A}} M_\alpha \right) \cap M_\gamma$ which indicates that $\cap_{\alpha \in \mathscr{A}} M_\alpha$ is not a minimal finite intersection. Hence, equality must hold.

$5 \Rightarrow 6$. If N is a submodule of M, we need to show that M/N is τ–cofinitely generated. Let $\{ M_\beta/N \}_{\beta \in \mathscr{B}}$ be the family of all τ–pure submodules of M/N. Then $t_\tau(M/N) = \cap_{\beta \in \mathscr{B}}(M_\beta/N) = \left(\cap_{\beta \in \mathscr{B}} M_\beta \right)/N$. By 4 there is a finite subset $\mathscr{A} \subseteq \mathscr{B}$ such that $\cap_{\alpha \in \mathscr{A}} M_\alpha = \cap_{\beta \in \mathscr{B}} M_\beta$. Thus, $t_\tau(M/N) = \left(\cap_{\alpha \in \mathscr{A}} M_\alpha \right)/N = \cap_{\alpha \in \mathscr{A}}(M_\alpha/N)$. Therefore, M/N is τ–cofinitely generated.

$6 \Rightarrow 1$. Suppose that every τ–torsionfree homomorphic image of M is τ–cofinitely generated and let $M_1 \supseteq M_2 \supseteq \cdots \supseteq M_k \supseteq \cdots$ be a decreasing chain of τ–pure submodules of M. Then $N = \cap_{k=1}^{\infty} M_k$ is an element of $\mathscr{P}_\tau(M)$ and so $M/N \in \mathscr{F}$. Thus, $0 = t_\tau(M/N) = \cap_{k=1}^{\infty} (M_k /N)$. But M/N is τ–cofinitely generated and so there is a positive integer n such that $0 = \cap_{k=1}^{n} (M_k /N) = M_n/N$. Thus, $M_n = N$ and so $M_n \supseteq M_k \supseteq N$ gives $M_k = M_n$ for all $k \geq n$. Therefore, M is τ–artinian. □

Corollary 2.4.4 *If* R *is* τ*–artinian and* I *is a right ideal of* R*, there is a positive integer* n *such that* $I^k/I^{k+1} \in \mathscr{T}$ *for all* $k \geq n$.

Proof. $I \supseteq I^2 \supseteq \cdots \supseteq I^k \supseteq \cdots$ is a decreasing chain of right ideals of R and so by the proposition there is a positive integer n such that $I^k/I^{k+1} \in \mathscr{T}$ for all $k \geq n$. □

Corollary 2.4.5 *If* R *is* τ*–artinian, there is a positive integer* n *such that* $J_\tau(R)^k / J_\tau(R)^{k+1} \in \mathscr{T}$ *for all* $k \geq n$.

When τ is the torsion theory in which every module is torsionfree, $J_\tau(R) = J(R)$ and if R is τ–artinian, Corollary 2.4.5 gives $J(R)^k = J(R)^{k+1} = \cdots$ for all $k \geq n$. In this case, R is right artinian and J(R) is nilpotent and so there is a positive integer n such that $J(R)^n = 0$. In fact, n can be selected to be the smallest positive integer for which $J(R)^n = J(R)^{n+1} = \cdots$

Proposition 2.4.6 *If* N *is a submodule of an* R*–module* M*, then* M *is* τ*–artinian if and only if both* N *and* M/N *are* τ*–artinian.*

Proof. First assume that M is τ–artinian. If $X/N \in \mathscr{P}_\tau(M/N)$, then $X \in \mathscr{P}_\tau(M)$. Consequently, any descending chain of modules in $\mathscr{P}_\tau(M/N)$ gives rise to a descending chain of modules in $\mathscr{P}_\tau(M)$ which will stabilize. Thus, the original chain will stabilize and so M/N is τ–artinian. Next, consider the mapping $\mathscr{P}_\tau(N) \to \mathscr{P}_\tau(M) : X \to X^c$ where the τ–pure closure of X is taken with respect to M. We have seen in the proof of Proposition 2.3.5 that this mapping is injective and so N is τ–artinian.

Conversely, suppose that both N and M/N are τ–artinian and let $M_1 \supseteq M_2 \supseteq \cdots \supseteq M_k \supseteq \cdots$ be a descending chain in $\mathscr{P}_\tau(M)$. First, note that $N/(N \cap M_k) \cong (N + M_k)/M_k \in \mathscr{F}$ and so $N \cap M_k$ is τ–pure in N. Hence,

$N \cap M_1 \supseteq N \cap M_2 \supseteq \cdots \supseteq N \cap M_k \supseteq \cdots$ is a descending chain in $\mathscr{P}_\tau(N)$ and so there is a positive integer m such that $N \cap M_k = N \cap M_m$ for all $k \geq m$. If $(N + M_k)^c$ is the τ–pure closure of $N + M_k$ in M for $i = 1, 2, \cdots$, then $(N + M_1)^c/N \supseteq (N + M_2)^c/N \supseteq \cdots (N + M_k)^c/N \supseteq \cdots$ is a descending chain in $\mathscr{P}_\tau(M/N)$. But M/N is τ–artinian and so there is a positive integer n such that $(N + M_k)^c/N = (N + M_n)^c/N$ for all $k \geq n$. It now follows that $(N + M_n)/(N + M_k)$ is isomorphic to a submodule of $(N + M_k)^c/(N + M_k) \in \mathscr{T}$ and so $(N + M_n)/(N + M_k) \in \mathscr{T}$ whenever $k \geq n$. If $k_0 = \max(m, n)$, then $(N + M_{k_0})/(N + M_k) \cong ((N + M_{k_0})/N)/((N + M_k)/N) \cong$ $(M_{k_0}/(N \cap M_{k_0}))/(M_k/(N \cap M_k)) \cong (M_{k_0}/(N \cap M_{k_0}))/(M_k/(N \cap M_{k_0})) \cong$ $M_{k_0}/M_k \in \mathscr{F}$. Hence, $M_{k_0}/M_k \in \mathscr{T} \cap \mathscr{F} = 0$ and so it must be the case that $M_k = M_{k_0}$ when $k \geq k_0$. □

Corollary 2.4.7 *If* $0 \to L \to M \to N \to 0$ *is a short exact sequence of* R–*modules, then* M *is* τ–*artinian if and only if* L *and* N *are* τ–*artinian.*

If $M \in \mathscr{T}$ or if M is τ–simple, $\mathscr{P}_\tau(M) = \{M\}$ or $\mathscr{P}_\tau(M) = \{0, M\}$, respectively, and so every descending chain in $\mathscr{P}_\tau(M)$ clearly stabilizes. Thus, M is τ–artinian. The following corollary to Proposition 2.4.6 is now immediate.

Corollary 2.4.8 *The following are equivalent for an* R–*module* M.

1. M *is* τ–*artinian.*
2. M/N *is* τ–*artinian for every* $N \in \mathscr{T}$.
3. $M/t(M)$ *is* τ–*artinian.*

Corollary 2.4.9 *Any finite direct product* $\prod_{i=1}^{n} M_i$ *of* R*–modules is* τ*–artinian if and only if* M_i *is* τ*–artinian for* $i = 1, 2, \cdots, n$.

Proof. Use induction. □

Corollary 2.4.10 *If* N *is a* τ*–artinian submodule of an* R*–module M, then* N^c *is* τ*–artinian.*

Proof. Since $N^c/N \in \mathcal{T}$, N^c/N is τ–artinian by the observation immediately preceding Corollary 2.4.8 that any module in $\mathcal{T}$ is τ–artinian. Hence, N^c is τ–artinian since N and N^c/N are τ–artinian. □

Proposition 2.4.11 R *is* τ*–artinian if and only if every* τ*–finitely generated* R*–module is* τ*–artinian.*

Proof. The proof of this proposition is essentially the same as the proof of Proposition 2.3.10 with minor modifications. □

§3 COMPOSITION SERIES RELATIVE TO TORSION THEORY THE GENERALIZED HOPKINS–LEVITZKI THEOREM

Hopkins [24] and Levitzki [27] have shown that every right artinian ring is right noetherian. The main result of this section is The Generalized Hopkins–Levitzki Theorem of Miller and Teply [30]. Following Goldman [23] we define a composition series of an R–module M relative to a torsion theory τ as follows:

Definition 3.1.1 A τ–*composition series* for an R–module M is a chain $0 = M_0 \subseteq M_1 \subseteq \cdots \subseteq M_n = M$ of submodules of M such that M_{k+1}/M_k is τ–simple for $k = 0, 1, 2, \cdots, n-1$. n is referred to as the *length of the* τ–*composition series* and if M has a τ–composition series, we say that M has τ–*finite length.*

As we shall soon see, if M has a τ–composition series of length n, then every τ–composition series of M has length n.

Proposition 3.1.2 *Any* R–*module which has* τ–*finite length is in* $\mathscr{F}$.

Proof. If $n = 1$, M is τ–simple and so $M \in \mathscr{F}$. Next, suppose that any R–module with τ–composition series of length k is in $\mathscr{F}$. If $0 = M_0 \subseteq M_1 \subseteq \cdots \subseteq M_{k+1} = M$ is a composition series for M, then $0 = M_1/M_1 \subseteq M_2/M_1 \subseteq \cdots \subseteq M_{k+1}/M_1 = M/M_1$ is a τ–composition series of M/M_1 of length k. Hence, $M/M_1 \in \mathscr{F}$. Therefore, the short exact sequence $0 \to M_1 \to M \to M/M_1 \to 0$ shows that $M \in \mathscr{F}$ since $\mathscr{F}$ is closed under extensions. $\square$

Proposition 3.1.3 *The following are equivalent for any* R–*module* M.

1. M *has* τ–*finite length.*
2. $M \in \mathscr{F}$ *and* M *is* τ–*noetherian and* τ–*artinian.*

Proof. $1 \Rightarrow 2$. If M has a τ–composition series of length 1, $\mathscr{P}_\tau(M) = \{0, M\}$ and so M is clearly τ–noetherian and τ–artinian. Next, suppose that any module with a τ–composition series of length k is τ–noetherian and τ–artinian and let $0 = M_0 \subseteq M_1 \subseteq \cdots \subseteq M_{k+1} = M$ be a τ–composition series of M of length $k+1$. Then $0 = M_1/M_1 \subseteq M_2/M_1 \subseteq \cdots \subseteq M_{k+1}/M_1 = M/M_1$ is a τ–composition series of M/M_1 of length k and so M/M_1 is both τ–noetherian and τ–artinian. Consequently, M is τ–noetherian and τ–artinian by Proposition 2.3.5 and Proposition 2.4.6, respectively. That $M \in \mathscr{F}$ follows from Proposition 3.1.2.

$2 \Rightarrow 1$. Since M is τ–artinian, use 4 of Proposition 2.4.3 to choose M_1 minimal among the nonzero the τ–pure submodules of M. Since $M \in \mathscr{F}$, Proposition 2.1.8 shows that M_1 is τ–simple. If $M_1 = M$, we are finished. If $M_1 \neq M$, construct a sequence $0 = M_0 \subseteq M_1 \subseteq \cdots \subseteq M_k \subseteq \cdots$ of submodules of M as follows: having chosen M_k, if $M_k \neq M$ choose M_{k+1} to be minimal among the τ–pure submodules of M which properly contain M_k. Then M_{k+1}/M_k is τ–simple. Since M is τ–noetherian, 4 of Proposition 2.3.3 shows that there is a maximal element M_n among the M_k 's. Clearly, $M_n = M$ for if not, our procedure would lead to an M_{n+1}. Therefore, we have constructed a τ–composition series for M and so M is of τ–finite length. □

Corollary 3.1.4 *If* M *has* τ–*finite length and* N *is a submodule of* M, *then* N *is of* τ–*finite length. Moreover, if* $N \in \mathscr{P}_\tau(M)$, M/N *has* τ–*finite length.*

Proof. Propositions 2.3.5 and 2.4.6 and the proposition above. □

Corollary 3.1.5 *If* $M \in \mathcal{F}$ *and* $N \in F_\tau(M)$ *has τ–finite length, then* M *has τ–finite length.*

Proof. Let $N \in F_\tau(M)$ be of τ–finite length and set $\mathcal{P}_\tau(M)^\# = \{ X \cap N \mid X \in \mathcal{P}_\tau(M) \}$. Since $N/(X \cap N) \cong (N + X)/X \subseteq M/X$, $N/(X \cap N) \in \mathcal{F}$ and so $X \cap N \in \mathcal{P}_\tau(N)$. But N is τ–artinian and τ–noetherian and so Propositions 2.4.3 and 2.3.3 show that there are elements X_1 and X_2 of $\mathcal{P}_\tau(M)$ such that $X_1 \cap N$ and $X_2 \cap N$ are minimal and maximal elements of $\mathcal{P}_\tau(N)$, respectively. We claim X_1 is a minimal element and X_2 is a maximal element of $\mathcal{P}_\tau(M)$. If $Y \subseteq X_1$, where $Y \in \mathcal{P}_\tau(M)$, then $Y \cap N = X_1 \cap N$ because of the minimality of $X_1 \cap N$. Since $X_1/Y \subseteq M/Y$, $X_1/Y \in \mathcal{F}$. But $X_1/Y \cong (X_1/(X_1 \cap N))/(Y/(X_1 \cap N))$ and $X_1/(X_1 \cap N) \cong (X_1 + N)/N \in \mathcal{T}$ since $N \in F_\tau(M)$. Hence, $X_1/Y \in \mathcal{T} \cap \mathcal{F} = 0$ and so $Y = X_1$. Similarly, X_2 is maximal in $\mathcal{P}_\tau(M)$. This suffices to show that M has τ–finite length. □

Corollary 3.1.6 *If* $0 \to L \to M \to N \to 0$ *is an exact sequence of* R*–modules such that* L *and* N *have τ–finite length, then* M *has τ–finite length.*

Proof. Since L and M are in $\mathcal{F}$, $M \in \mathcal{F}$ since $\mathcal{F}$ is closed under extensions. Also L and N are both τ–noetherian and τ–artinian and so it follows from Corollary 2.3.6 and Corollary 2.4.7 that M is τ–noetherian and τ–artinian. Hence, M has τ–finite length by the proposition above. □

Proposition 3.1.7 *If an R–module M has τ–finite length, then any two τ–composition series of M have the same length.*

Proof. If an R–module M has a τ–composition series, then clearly among all the τ–composition series for M there must be one of minimal length. Our proof is by induction on the length of this minimal τ–composition series. Suppose M has a τ–composition series $0 \subseteq M_1 = M$ of length 1 and a τ–composition series $0 = N_0 \subseteq N_1 \subseteq \cdots \subseteq N_n = M$ of length n, then M is τ–simple and so $M/N_{n-1} \in \mathscr{T}$. But M/N_{n-1} is τ–simple and so $M/N_{n-1} \in \mathscr{F}$. Thus, $M/N_{n-1} = 0$, a contradiction since a τ–simple R–module must be nonzero. Thus, if an R–module M has a τ–composition series of minimal length 1, then every τ–composition series of M must be of length 1. Now assume that if an R–module M has a τ–composition series of minimal length k, then every τ–composition series of M must be of length k. Next, suppose that M has a τ–composition series $0 = N_0 \subseteq N_1 \subseteq \cdots \subseteq N_{k+1} = M$ of minimal length $k + 1$. If we set $N = N_k$, then $0 = N_0 \subseteq N_1 \subseteq \cdots \subseteq N_k = N$ is a τ–composition series of N which is of minimal length k since if not, we could construct a τ–composition series of M of length less than $k + 1$. Finally, let $0 = M_0 \subseteq M_1 \subseteq \cdots \subseteq M_n = M$ be a τ–composition series of M of length n, $n \geq k + 1$. The proof will be complete if we can show that $n = k + 1$. Since $0 = M_0 \subseteq N$, let m be the largest integer such that $M_m \subseteq N$. We claim that $0 = M_0 \subseteq M_1 \subseteq \cdots \subseteq M_m \subseteq M_{m+1} \cap N \subseteq \cdots \subseteq M_n \cap N = N$ is a τ–composition series of N of length $n - 1$. Clearly, M_{k+1}/M_k is τ–simple for $0 \leq k \leq m - 1$. Next, we claim that $(M_{k+1} \cap N)/(M_k \cap N)$ is τ–simple for $m + 1 \leq k \leq n - 1$. Notice that $(M_{k+1} \cap N)/(M_k \cap N) =$ $(M_{k+1} \cap N)/(M_{k+1} \cap N \cap M_k) \cong [(M_{k+1} \cap N) + M_k]/M_k \subseteq M_{k+1}/M_k$ and so $(M_{k+1} \cap N)/(M_k \cap N)$ is either zero or τ–simple since nonzero submodules of τ–simple modules are τ–simple. If $(M_{k+1} \cap N)/(M_k \cap N) = 0$, then $M_k \cap N = M_{k+1} \cap N$. Hence, $M_k/(M_k \cap N) \subseteq M_{k+1}/(M_{k+1} \cap N)$

and $M_{k+1}/(M_{k+1} \cap N) \cong (M_{k+1} + N)/N \subseteq M/N$. But M/N is τ–simple and so if $M_k/(M_k \cap N) \neq 0$, $(M/N)/(M_k/(M_k \cap N)) \in \mathcal{T}$. Now M_{k+1}/M_k is isomorphic to a submodule of $(M/N)/(M_k/(M_k \cap N))$ and so it must be the case that $M_{k+1}/M_k = 0$, a contradiction because M_{k+1}/M_k is τ–simple. Thus, $M_k/(M_k \cap N) = 0$. Hence, $(M_{k+1} \cap N)/(M_k \cap N) = 0$ implies that$M_k/(M_k \cap N) = 0$ which in turn tells us that $M_k \subseteq N$. But $k \geq m + 1$ so this contradicts the maximality of m. Therefore, $(M_{k+1} \cap N)/(M_k \cap N)$ is nonzero and so is τ–simple for $m + 1 \leq k \leq n - 1$.

Next, we will show that the length of that this τ–composition series of N is $n - 1$ by showing that $M_m = M_{m+1} \cap N$. If $(M_{m+1} \cap N)/M_m \neq 0$, $(M_{m+1} \cap N)/M_m \subseteq M_{m+1}/M_m$ implies that $M_{m+1}/(M_{m+1} \cap N) \in \mathcal{T}$. But $M_{m+1}/(M_{m+1} \cap N) \cong (M_{m+1} + N)/N$ implies that $M_{m+1}/(M_{m+1} \cap N)$ is in $\mathcal{F}$ since $(M_{m+1} + N)/N \subseteq M/N$ and M/N is τ–simple. Hence, $M_{m+1}/(M_{m+1} \cap N) = 0$ which shows that $M_{m+1} = M_{m+1} \cap N$. But then $M_{m+1} \subseteq N$ which contradicts the maximality of m. Thus, $M_m = M_{m+1} \cap N$ as was indicated. Consequently, $0 = M_0 \subseteq M_1 \subseteq \cdots \subseteq M_m \subseteq M_{m+1} \cap N \subseteq \cdots \subseteq M_n \cap N = N$ is a composition series of N of length $n - 1$. Moreover, we have seen that N is an R–module with a τ–composition series of minimal length k. Hence, by the induction hypothesis, $n - 1 = k$ and so $n = k + 1$. □

The following work is due to Miller and Teply [30]. Terminology has been changed, however.

Definition 3.1.8 An R–module M is said to be *finitely τ–radical free,* if there exists a finite set $\{N_i\}_{i=1}^{n}$ of τ–maximal submodules of M such that $J_\tau(M) = \cap_{i=1}^{n} N_i = 0$.

Proposition 3.1.9 *Any finitely τ–radical free* R*–module has τ–finite length.*

Proof. If M is a finitely τ–radical free R–module, Proposition shows that M embeds in a finite direct product of τ–simple R–modules. Since this direct product clearly has τ–finite length, Corollary 3.1.4 shows that M has τ–finite length. □

Proposition 3.1.10 *If* M *is* τ*–torsionfree, then any* τ*–dense submodule of* M *is essential in* M.

Proof. Let $M \in \mathscr{F}$ and suppose $N \in F_\tau(M)$. If $m \in M \setminus N$, then $(mR + N)/N \subseteq M/N \in \mathscr{T}$. But $(mR + N)/N \cong mR/(mR \cap N)$ and so if $mR \cap N = 0$, we would have a contradiction since mR is τ–torsionfree and $m \neq 0$. □

Proposition 3.1.11 *If* M *is a finitely* τ*–radical free* R*–module, there are* τ*–simple submodules* $N_1, N_2, \cdots, N_n$ of M *such that the sum* $\sum_{i=1}^{n} N_i$ *is direct and such that* $\sum_{i=1}^{n} N_i$ *is* τ*–essential in* M.

Proof. If M is finitely τ–radical free, M has τ–finite length and so Proposition 3.1.3 shows that M is τ–artinian. Hence, there exists a minimal finite family Ω of τ–maximal submodules of M such that $J_\tau(M) = \cap_{X \in \Omega} X = 0$. Suppose Card $\Omega = n$. We choose the τ–simple submodules $\{ N_i \}_{i=1}^{n}$ of M according to the following procedure: Since M is τ–artinian and $M \in \mathscr{F}$, let N_1 be minimal among the nonzero modules in $\mathscr{P}_\tau(M)$. By Proposition 2.1.8 N_1 is τ–simple. If $0 \neq x_1 \in N_1$, then $x_1 \notin J_\tau(M) = 0$ and so there is a τ–maximal submodule $X_1 \in \Omega$ such that $x_1 \notin X_1$. Now $0 \neq N_1/(N_1 \cap X_1) \cong (N_1 + X_1)/X_1 \in \mathscr{F}$ since $X_1 \in \mathscr{P}_\tau(M)$. But N_1 is τ–simple and so if $N_1 \cap X_1 \neq 0$, $N_1/(N_1 \cap X_1) \in \mathscr{T}$ in which case $N_1/(N_1 \cap X_1) = 0$. But then $N_1 = N_1 \cap X_1$ and so $N_1 \subseteq X_1$ which contradicts the choice of X_1. Hence, $N_1 \cap X_1 = 0$. Next, choose N_2 to be minimal among the nonzero τ–pure submodules of X_1, then as before N_2 is τ–simple. If $0 \neq x_2 \in N_2$, there is an $X_2 \in \Omega$ which is distinct from X_1 such that $x_2 \notin X_2$ and $N_2 \cap X_2 = 0$. Similarly, there is a τ–simple $N_3 \subseteq X_1 \cap X_2 \neq 0$ and an $X_3 \in \Omega$ which is distinct from X_1 and X_2 such that $N_3 \cap X_3 = 0$. Finally, choose N_n to be a τ–simple submodule of $\cap_{i=1}^{n-1} X_i \neq 0$ and X_n the last τ–maximal submodule in Ω which is distinct from X_i for $i = 1, 2, \cdots, n-1$ and $N_n \cap X_n = 0$. $\sum_{i=1}^{n} N_i$ is direct by construction and so it remains only to show that $\sum_{i=1}^{n} N_i$ is τ–essential in M. We will first show that $M/\left(\sum_{i=1}^{k} N_i + \cap_{i=1}^{k} X_i\right) \in \mathcal{T}$ for $k = 1, 2, \cdots, n$. Since M/X_1 is τ–simple and $(N_1 + X_1)/X_1 \neq 0$, $M/(N_1 + X_1) \in \mathcal{T}$. Thus, the case for $k = 1$ is established. Now assume $M/\left(\sum_{i=1}^{k} N_i + \cap_{i=1}^{k} X_i\right) \in \mathcal{T}$ for all $k < m$. Since $0 \neq (N_m + X_m)/X_m \subseteq \left(\cap_{i=1}^{m-1} X_i + X_m\right)/X_m$ and M/X_m is τ–simple, $\left(\cap_{i=1}^{m-1} X_k + X_m\right)/(N_m + X_m) \in \mathcal{T}$. Also since $N_m \subseteq \cap_{i=1}^{m-1} X_i$, for each $x \in \cap_{i=1}^{m-1} X_i$, $\left(N_m + \cap_{i=1}^{m} X_i : x\right) = (N_m + X_m : x)$. Thus, it follows from $\left(\cap_{i=1}^{m-1} X_i + X_m\right)/(N_m + X_m) \in \mathcal{T}$ that $\cap_{i=1}^{m-1} X_i /\left(N_m + \cap_{i=1}^{m} X_i\right) \in \mathcal{T}$. But, by hypothesis, $\left(\sum_{i=1}^{m-1} N_i + \cap_{i=1}^{m-1} X_i\right) \in \mathcal{T}$ and so we see that $\left(\sum_{i=1}^{m-1} N_i + \cap_{i=1}^{m-1} X_i\right)/\left(\sum_{i=1}^{m} N_i + \cap_{i=1}^{m} X_i\right) \in \mathcal{T}$ since $\mathcal{T}$ is closed under homomorphic images. Therefore, the short exact sequence

$$0 \to \frac{\sum_{i=1}^{m-1} N_i + \cap_{i=1}^{m-1} X_i}{\sum_{i=1}^{m} N_i + \cap_{i=1}^{m} X_i} \to \frac{M}{\sum_{i=1}^{m} N_i + \cap_{i=1}^{m} X_i} \to \frac{M}{\sum_{i=1}^{m-1} N_i + \cap_{i=1}^{m-1} X_i} \to 0$$

tells us that $\dfrac{M}{\sum_{i=1}^{m} N_i + \cap_{i=1}^{m} X_i} \in \mathcal{T}$ since $\mathcal{T}$ is closed under extensions.

Consequently, by induction, $\dfrac{M}{\sum_{i=1}^{k} N_i + \cap_{i=1}^{k} X_i} \in \mathcal{T}$ for $k = 1, 2, \cdots, n$.

But when $k = n$, $\dfrac{M}{\sum_{i=1}^{n} N_i + \cap_{i=1}^{n} X_i} \in \mathcal{T}$ becomes $\dfrac{M}{\sum_{i=1}^{n} N_i}$ since

$\cap_{i=1}^{n} X_i = 0$. Hence, $\sum_{i=1}^{n} N_i \in F_\tau(M)$. It now follows from Proposition 3.1.10 that $\sum_{i=1}^{n} N_i$ is τ–essential in M. □

Corollary 3.1.12 *If* M *is a* τ*–radical free and* τ*–artinian* R*–module, then there exist* τ*–simple submodules* $N_1, N_2, \cdots, N_n$ *of* M *such that the sum* $\sum_{i=1}^{n} N_i$ *is direct and* $\sum_{i=1}^{n} N_i$ *is a* τ*–essential submodule of* M.

We conclude this section with The Generalized Hopkins–Levitzki Theorem of Miller and Teply [30].

Proposition 3.1.13 (**The Generalized Hopkins–Levitzki Theorem**) *If* R *is* τ*–artinian, then any* τ*–artinian* R*–module is* τ*–noetherian. In particular,* R *is* τ*–noetherian.*

Proof. Suppose that M is τ–artinian. If $M = t(M)$, then M is clearly τ–noetherian and we are finished. If $M \neq t(M)$, construct an ascending chain of modules in $\mathscr{P}_\tau(M)$ as follows: Let $M_0 = t(M)$ and choose $M_1 \subseteq M$ to be such that M_1/M_0 is minimal in the set of nonzero τ–pure submodules of M/M_0. Such an M_1/M_0 exists since, by Proposition 2.4.6, M/M_0 is τ–artinian. Clearly, $M_1 \in \mathscr{P}_\tau(M)$. By Proposition 2.1.8, M_1/M_0 is τ–simple and so is τ–noetherian. Thus, by Proposition 2.3.5, M_1 is τ–noetherian. If $M = M_1$, M is τ–noetherian and the proof is complete. If $M \neq M_1$, the process can be repeated to choose an $M_2 \in \mathscr{P}_\tau(M)$ such that M_2/M_1 is τ–simple and M_2 is τ–noetherian. If this process does not yield a positive integer k for which M_k is τ–noetherian and $M = M_k$, we can inductively construct a strictly increasing chain of modules $M_0 \subset M_1 \subset \cdots \subset M_k \subset \cdots$ in $\mathscr{P}_\tau(M)$ such that for $k \geq 1$, M_k/M_{k-1} is τ–simple and M_k is τ–noetherian. We now show that this is impossible: Suppose $M \neq M_k$ for each $k \geq 1$ and set

$$m_0 = 0 \quad \text{and for } t \geq 1,$$

$$m_{t+1} = \max \Gamma_t \quad \text{where } \Gamma_t = \{\, k \geq 1 \mid xJ_\tau(R) \subseteq M_{m_t} \text{ for some } x \in M_k \setminus M_{k-1} \,\}.$$

Notice that $m_t + 1 \in \Gamma_t$, since M_{m_t+1}/M_{m_t} is τ–simple and $(M_{m_t+1}/M_{m_t})J_\tau(R) = 0$. Assume that m_t exists and let's show that m_{t+1} exists. If m_{t+1} does not exist, there is an infinite set Ω of indices $k > m_t + 1$ for which we may choose $x_k \in M_k \setminus M_{k-1}$ such that $x_k J_\tau(R) \subseteq M_{m_t}$. Since R is τ–artinian, $R/J_\tau(R)$ is τ–radical free and τ–artinian as an R–module and so, by Corollary 3.1.12, there is a set $K_1/J_\tau(R)$, $K_2/J_\tau(R)$, $\cdots$, $K_n/J_\tau(R)$ of τ–simple R–submodules of $R/J_\tau(R)$ such that the sum $\sum_{i=1}^{n}(K_i/J_\tau(R))$ is direct and such that $\sum_{i=1}^{n}(K_i/J_\tau(R))$ is τ–essential in $R/J_\tau(R)$. If for each $i = 1, 2, \cdots, n$, $x_k K_i \subseteq M_{m_t}$, then $x_k \sum_{i=1}^{n} K_i \subseteq M_{m_t}$. Now $\sum_{i=1}^{n} K_i \in F_\tau(R)$ and so we see that $x_k + M_{m_t}$ is a nonzero τ–torsion element of M/M_{m_t}. By construction, $M/M_{m_t} \in \mathscr{F}$ and so we have a contradiction. Therefore, for at least one of the $K_i/J_\tau(R)$, $x_k K_i \not\subseteq M_{m_t}$.

Now assume that for any $K_i/J_\tau(R)$ such that $x_k K_i \not\subseteq M_{m_t}$, we have that M_{m_t} is properly contained in $(x_k K_i + M_{m_t}) \cap M_{k-1}$. Since $(x_k K_i + M_{m_t})/M_{m_t} \subseteq M/M_{m_t} \in \mathscr{F}$ and $K_i/J_\tau(R)$ is τ–simple, the canonical mapping $K_i/J_\tau(R) \to (x_k K_i + M_{m_t})/M_{m_t}$ is an isomorphism. Thus, $(x_k K_i + M_{m_t})/M_{m_t}$ is τ–simple which leads to $(x_k K_i + M_{m_t})/((x_k K_i + M_{m_t}) \cap M_{k-1}) \cong ((x_k K_i + M_{m_t})/M_{m_t})/(((x_k K_i + M_{m_t}) \cap M_{k-1})/M_{m_t}) \in \mathscr{T}$. But $(x_k K_i + M_{m_t})/((x_k K_i + M_{m_t}) \cap M_{k-1}) \cong (x_k K_i + M_{k-1})/M_{k-1} \subseteq M/M_{k-1} \in \mathscr{F}$. Hence, $(x_k K_i + M_{m_t})/((x_k K_i + M_{m_t}) \cap M_{k-1}) = 0$ and so we arrive at $x_k K_i \subseteq M_{k-1}$ for any $K_i/J_\tau(R)$ with $x_k K_i \not\subseteq M_{m_t}$. But then $x_k \sum_{i=1}^{n} K_i \subseteq M_{k-1}$, a contradiction since $x_k \notin M_{k-1}$, $\sum_{i=1}^{n} K_i \in F_\tau(R)$ and $M/M_{k-1} \in \mathscr{F}$.

We conclude that for each $k \in \Omega$, there is a right ideal K_k of R and an $x_k \in M_k \setminus M_{k-1}$ such that $x_k K_k \not\subseteq M_{m_t}$, $(x_k K_k + M_{m_t})/M_{m_t}$ is τ–simple and $(x_k K_k + M_{m_t}) \cap M_{k-1} = M_{m_t}$. It is not difficult to show that the sum $\sum_{k \in \Omega}[(x_k K_k + M_{m_t})/M_{m_t}] \subseteq M/M_{m_t}$ is direct. Next, let $\Omega = \{k_1, k_2, \cdots\}$ and set $N_1 = M$ and for $\alpha > 1$ choose N_α maximal with respect to the property $N_\alpha \supseteq \sum_{i=\alpha}^{\infty}(x_{k_i}K_{k_i} + M_{m_t})$, $N_\alpha \subseteq N_{\alpha-1}$ and $N_\alpha \cap \sum_{i=1}^{\alpha-1}(x_{k_i}K_{k_i} + M_{m_t}) = M_{m_t}$. Then $M = N_1 \supseteq N_2 \supseteq \cdots \supseteq N_\alpha \supseteq \cdots$ is a strictly decreasing chain of submodules in $\mathscr{P}_\tau(M)$ which is a contradiction since M is τ–artinian. Consequently, m_{t+1} must exist.

Since M_{m_t+1}/M_{m_t} is τ–simple and $(M_{m_t+1}/M_{m_t})J_\tau(R) = 0$, we see that $m_{t+1} \geq m_t + 1 > m_t$ for all $t > 0$. Hence, the infinite sequence $\{m_t\}_{t=1}^{\infty}$ is strictly increasing. By Corollary 2.4.5, there is a positive integer n such that $J_\tau(R)^{n+q}/J_\tau(R)^{n+q+1} \in \mathscr{T}$ for all $q \geq 0$, so let $x \in M_{m_n+1} \setminus M_{m_n}$. Then $x J_\tau(R) \not\subseteq M_{m_{n-1}}$ and so $xa_0 \notin M_{m_{n-1}}$ for some $a_0 \in J_\tau(R)$. But $xa_0 J_\tau(R) \not\subseteq M_{m_{n-2}}$ and so $xa_0a_1 \notin M_{m_{n-2}}$ for some $a_1 \in J_\tau(R)$. Inductively we obtain $a_0, a_1, \cdots, a_{n-1} \in J_\tau(R)$ such that $xa_0a_1a_2 \cdots a_i J_\tau(R) \not\subseteq M_{m_{n-1-i}}$ for $i = 0, 1, \cdots, n-1$. In particular, $xa_0a_1a_2 \cdots a_{n-1} J_\tau(R) \not\subseteq M_{m_0} = M_0$ and so $x J_\tau(R)^n \not\subseteq M_0$. But $x \in M_{m_n+1}$ and so $x J_\tau(R)^{m_n+1} \subseteq M_0$ since M_k/M_{k-1} is τ–simple for all $k \geq 1$. Consequently, there is an integer $q \geq n$ for which $x J_\tau(R)^q \not\subseteq M_0$ and $x J_\tau(R)^{q+1} \subseteq M_0$.

Now $\varphi : R/J_\tau(R)^{q+1} \to (xR + M_0)/M_0 : r + J_\tau(R)^{q+1} \to xr + M_0$ is an epimorphism and $0 \neq \varphi(J_\tau(R)^q/J_\tau(R)^{q+1}) \subseteq M/M_0 = M/t(M) \in \mathscr{F}$. But since $q \geq n$, $\varphi(J_\tau(R)^q/J_\tau(R)^{q+1}) \in \mathscr{T}$ which contradicts $\mathscr{T} \cap \mathscr{F} = 0$. Therefore, the chain $M_0 \subset M_1 \subset \cdots \subset M_k \subset \cdots$ cannot be strictly increasing and so $M = M_k$ for some $k \geq 1$. Consequently, M must be τ–noetherian. □

§4 INJECTIVE AND PROJECTIVE CONCEPTS AND TORSION THEORY THE GENERALIZED BAER AND FUCHS CONDITIONS RELATIVELY FLAT MODULES

4.1 (Quasi–)Injective Modules Relative to a Torsion Theory

Definition 4.1.1 An R–module M is said to be to be τ–*injective* if every row exact diagram of R–modules and R–module homomorphisms

$$\begin{array}{ccccccccc} 0 & \rightarrow & L & \rightarrow & X & \rightarrow & N & \rightarrow & 0 \\ & & \downarrow & & & & & & \\ & & M & & & & & & \end{array}$$

where $N \in \mathscr{T}$ can be completed commutatively by an R–linear map $X \rightarrow M$.

Observe that when τ is the torsion theory in which every module is torsion, an R–module is τ–injective if and only if it is injective.

It is easy to shown that an R–module M is τ–injective if and only if whenever $0 \rightarrow L \rightarrow X \rightarrow N \rightarrow 0$ is exact with $N \in \mathscr{T}$, the induced map $\mathrm{Hom}_R(X, M) \rightarrow \mathrm{Hom}_R(L, M) \rightarrow 0$ is an epimorphism. This in turn leads to that fact that M is τ–injective if and only if $\mathrm{Ext}^1_R(N, M) = 0$ for all $N \in \mathscr{T}$.

The Generalized Baer Condition

One well known result concerning injective modules states that an R–module M is injective if and only if for each (essential) right ideal K of R and every $f \in \mathrm{Hom}_R(K, M)$ there is an $m \in M$ such that $f(k) = mk$ for all

$k \in K$. This is known as *Baer's Condition* [1]. Since R has an identity, Baer's Condition is equivalent to being able to extend f linearly to R. Baer's result shows that the right ideals of R form a test set for injectivity. The following proposition relates the set of right ideals in $F_\tau(R)$ to τ–injectivity.

Proposition 4.1.2 (The Generalized Baer Condition) *An* R–*module* M *is* τ–*injective if and only if for each* $K \in F_\tau(R)$ *and each* $f \in \mathrm{Hom}_R(K, M)$, *there is an* $m \in M$ *such that* $f(k) = mk$ *for all* $k \in K$.

Proof. If M is τ–injective, $K \in F_\tau(R)$ and $f \in \mathrm{Hom}_R(K, M)$, the τ–injectivity of M implies that there is a $g \in \mathrm{Hom}_R(R, M)$ which extends f. If $m = g(1)$, then $f(k) = g(k) = g(1)k = mk$ for all $k \in K$.

Conversely, suppose the condition holds and consider the row exact diagram

$$\begin{array}{ccccc} 0 & \rightarrow & L & \xrightarrow{k} & N \\ & & f\downarrow & & \\ & & M & & \end{array}$$

where $L \in F_\tau(M)$ and $k(x) = x$ for all $x \in L$. Let $\mathscr{S}$ be the set of all (L', f') such that $L \subseteq L' \subseteq N$, $f' : L' \rightarrow M$ is R–linear and $f'|L = f$. Partial order $\mathscr{S}$ by $(L', f') \leq (L'', f'')$ if and only if $L' \subseteq L''$ and f'' induces f'. Via Zorn's Lemma it can be shown that $\mathscr{S}$ has a maximal element, say (N^*, f^*). If $N^* \neq N$, let $y \in N \setminus N^*$. Since $N^* \in F_\tau(N)$, $N/N^* \in \mathscr{T}$ and so $(N^*: y) = (0 : y + N^*) \in F_\tau(R)$. Hence, we have an R–linear mapping $g : (N^*: y) \rightarrow M : r \rightarrow f^*(yr)$ and so there is an $m \in M$ such that $g(r) = mr$ for all $r \in (N^*: y)$. Now define $\phi : N^* + yR \rightarrow M$ by $\phi(x + yr) = f^*(x) + mr$.

Then ϕ properly extends f^* which contradicts the maximality of (N^*, f^*). Therefore, $N^* = N$ and f^* completes the diagram commutatively. □

Corollary 4.1.3 *An* R*–module* M *is* τ*–injective if and only if for each* τ*–essential right ideal* E *of* R *and each* $f \in \mathrm{Hom}_R (E , M)$, *there is an* $m \in M$ *such that* $f(k) = mk$ *for all* $k \in E$.

Proof. Let $K \in F_\tau(R)$. Use Zorn's lemma and choose J maximal among the right ideals K′ of R such that $K \cap K' = 0$. Then $E = K \oplus J$ is an essential right ideal of R and $E \in F_\tau(R)$. If $f \in \mathrm{Hom}_R (K, M)$, $g : E \to M : x + y \to f(x)$ extends f to E. The corollary now follows from these observations and Proposition 4.1.2. □

In view of The Generalized Baer Condition, an R–module M is τ–injective if and only if $\mathrm{Ext}^1_R (R/K, M) = 0$ for all right ideals $K \in F_\tau(R)$ if and only if $\mathrm{Ext}^1_R (R/E, M) = 0$ for all τ–essential right ideals E of R.

Note that the direct product of a family of R–modules is τ–injective if and only if each factor is τ–injective. The standard argument for injectives works with necessary changes being made to accommodate the torsion theory.

Definition 4.1.4 An R–module M is said to be τ*–completely reducible* if every $N \in F_\tau(M)$ is a direct summand of M. R is τ–completely reducible if R is τ–completely reducible as an R–module.

Proposition 4.1.5 *The following are equivalent for a ring* R.

1. *Every* R*–module is* τ*–injective.*
2. *Every* R*–module in* $\mathscr{T}$ *is projective.*

3. *Every* R*–module is* τ*–completely reducible.*
4. *Every cyclic R–module in* $\mathscr{T}$ *is projective.*
5. R *is* τ*–completely reducible.*

Proof. $1 \Rightarrow 2$. Suppose $M \in \mathscr{T}$ and let $F \to M \to 0$ be a free module on M with kernel K. The sequence $0 \to K \to F \to M \to 0$ splits, since K is τ–injective and so M is isomorphic to a direct summand of F. Consequently, M is projective.

$2 \Rightarrow 4$ is obvious.

$4 \Rightarrow 5$. If $K \in F_\tau(R)$, then $0 \to K \to R \to R/K \to 0$ splits because R/K is cyclic and in $\mathscr{T}$. Hence, K is a direct summand of R.

$5 \Rightarrow 1$ Suppose $f \in \mathrm{Hom}_R(K, M)$ where $K \in F_\tau(R)$ and let K' be a right ideal of R such that $R = K \oplus K'$. Extend f to R by defining $f^* : R \to M$ to be the map $f^*(k + k') = f(k)$. If $f^*(1) = m$, then $f(k) = mk$ for all $k \in K$ and so M is τ–injective by The Generalized Baer Condition.

$2 \Rightarrow 3$. If $N \in F_\tau(M)$, then $0 \to N \to M \to M/N \to 0$ splits since M/N is projective. Thus, N is a direct summand of M. Hence, $2 \Rightarrow 3$. Since, 3 clearly implies 5, the proof is complete. □

Definition 4.1.6 An R–linear mapping $f : M \to \oplus_{\alpha \in \Delta} M_\alpha$ is said to be a *finite homomorphism* if $f(M)$ has nonzero coordinates in at most a finite number of the members of the family $\{ M_\alpha \}_{\alpha \in \Delta}$.

Proposition 4.1.7 *Let* τ *be a torsion theory on* Mod–R *and suppose that* $t_\tau(R) \subseteq K_1 \subseteq K_2 \subseteq K_3 \subseteq \ldots$ *is a countably infinite increasing chain of right ideals of* R. *If* $K = \cup_{i=1}^{\infty} K_i$ *and* $\{ M_\alpha \}_{\alpha \in \Delta}$ *is a family of* τ*–torsionfree* *injective R–modules, let* $f : K \to \oplus_{\alpha \in \Delta} M_\alpha$ *be an R–linear mapping such that* $f|_{K_i}$ *is a finite homomorphism for each* $i \geq 1$. *If there exists a positive integer* n *such that* $K/K_n \in \mathscr{T}$, *then* f *can be extended to an R–linear mapping* $g : R \to \oplus_{\alpha \in \Delta} M_\alpha$ *which is a finite homomorphism.*

Proof. Suppose such a positive integer n exists and consider the diagram

$$\begin{array}{cccc} 0 \to & K_n & \to & R \\ & \downarrow f|_{K_n} & \swarrow g & \\ & \oplus_{\alpha \in \Delta} M_\alpha & & \end{array}$$

Since $f|_{K_n}$ is a finite homomorphism, $f|_{K_n}$ is actually an R–linear mapping into a finite number of components of $\oplus_{\alpha \in \Delta} M_\alpha$. But since a finite direct sum of injective modules is injective, it follows that there is an R–linear mapping $g : R \to \oplus_{\alpha \in \Delta} M_\alpha$ which makes the diagram commute. Since both f and $g|_K$ extend $f|_{K_n}$, $K_n \subseteq \ker(f - g|_K)$. Therefore, $f - g|_K$ induces an R–linear mapping $\varphi : K/K_n \to \oplus_{\alpha \in \Delta} M_\alpha$. But since $K/K_n \in \mathscr{T}$ and $\oplus_{\alpha \in \Delta} M_\alpha \in \mathscr{F}$, it must be the case that $\varphi = 0$. Hence, $f = g|_K$. g is clearly a finite homomorphism since $g(1)$ has at most a finite number of nonzero components in $\oplus_{\alpha \in \Delta} M_\alpha$. □

Matlis [29] has shown that a ring R is right noetherian if and only if every direct sum of injective R–modules is injective. Matlis' result can be recovered from the following two propositions when τ is chosen to be the torsion theory in which every R–module is torsionfree. Both propositions appear in [36].

Proposition 4.1.8 *The following are equivalent for a torsion theory* τ *on* Mod–R.

1. *Every direct sum of* τ*–torsionfree injective* R*–modules is injective.*
2. *Every direct sum of countably many* τ*–torsionfree injective* R*–modules is injective.*
3. R *is* τ*–noetherian.*
4. *If* $K_1 \subset K_2 \subset K_3 \subset \ldots$ *is a strictly increasing countably infinite chain of right ideals of* R*, then* $\left(\cup_{i=1}^{\infty} K_i\right)/K_n \in \mathcal{T}$ *for some integer* $n \geq 1$.
5. *If* $t_\tau(R) \subset K_1 \subset K_2 \subset K_3 \subset \ldots$ *is a strictly increasing countably infinite chain of right ideals of* R*, then* $\left(\cup_{i=1}^{\infty} K_i\right)/K_n \in \mathcal{T}$ *for some integer* $n \geq 1$.

Proof. $1 \Rightarrow 2$ is obvious. $2 \Rightarrow 3$. Let $K_1 \subset K_2 \subset K_3 \subset \ldots$ be a strictly increasing chain of τ–pure right ideals of R. Then $K_i = (0 : 1 + K_i)$ and $(1 + K_i)R \cong R/K_i \in \mathcal{F}$ for each i. Hence, we can consider each K_i to be of the form $(0 : x_i)$ where $x_iR \in \mathcal{F}$. Since $x_iR \in \mathcal{F}$, $E(x_iR) \in \mathcal{F}$, so let $K = \cup_{i=1}^{\infty} K_i$ and $E = \oplus_{i=1}^{\infty} E(x_iR)$. Note that $x_iK \neq 0$ for each i for if such were not the case, the given sequence of τ–pure right ideals of R would not be strictly increasing. Now define $f : K \to E$ by $f(k) = (x_ik)$. Then by 2 and Baer's Condition there is a $z \in E$ such that $f(k) = zk$ for all $k \in K$. But it is possible for z to be nonzero in only finitely many summands of E and so it must be the case that f(K) can be nonzero only in the same summands of E. But from the definition of f we see that this contradicts the fact that $x_iK \neq 0$ for all i. Hence, the set of τ–pure right ideals of R must satisfy the ascending chain condition and so R is τ–noetherian.

$3 \Rightarrow 4$. Let $K_1 \subset K_2 \subset K_3 \subset \ldots$ be a strictly increasing countably infinite chain of right ideals of R and for each i choose a right ideal I_i of R containing K_i such that $I_i/K_i = t_\tau(R/K_i)$. Then $R/I_i \cong (R/K_i)/(I_i/K_i)$ is in $\mathcal{F}$. Hence, each I_i is a τ–pure right ideal of R and $I_1 \subseteq I_2 \subseteq I_3 \subseteq \cdots$. But R is τ–noetherian and so there is a positive integer n such that $I_n = I_{n+1} = \cdots$. Hence, $\cup_{i=1}^{\infty} K_i \subseteq \cup_{i=1}^{\infty} I_i = I_n$ and so $\left(\cup_{i=1}^{\infty} K_i\right)/K_n \subseteq I_n/K_n \in \mathcal{T}$.

$4 \Rightarrow 5$ is obvious.

$5 \Rightarrow 1$. Let K be a right ideal of R and suppose that $\{M_\alpha\}_{\alpha \in \Delta}$ is a family of τ–torsionfree injective R–modules. Furthermore, suppose that $f : K \to \oplus_{\alpha \in \Delta} M_\alpha$ is an R–linear mapping. By Baer's Condition, to show that $\oplus_{\alpha \in \Delta} M_\alpha$ is injective, it suffices to show that f can be extended to R. Since $\oplus_{\alpha \in \Delta} M_\alpha \in \mathcal{F}$, we see that $f(t_\tau(K)) = 0$ and so f can be extended to $K + t_\tau(R)$ by simply defining the extension to be zero on $t_\tau(R)$. Moreover, any extension of f to R, say $g : R \to \oplus_{\alpha \in \Delta} M_\alpha$, has the property that $g(t_\tau(R)) = 0$ and so we can assume, without loss of generality, that $t_\tau(R) \subseteq K$. Now define an increasing transfinite sequence of right ideals K_β of R such that $t_\tau(R) \subseteq K_\beta \subseteq K$ as follows:

i. Set $K_1 = t_\tau(R)$

ii. If $\beta > 1$ is not a limit ordinal and $K/K_{\beta-1}$ is finitely generated, set $K_\beta = K$.

iii. If $\beta > 1$ is not a limit ordinal and $K/K_{\beta-1}$ is not finitely generated, choose a countably infinite sequence $x_{\beta 1}, x_{\beta 2}, x_{\beta 3}, \ldots$ of elements of K such that $K_{\beta-1} \subset K_{\beta-1} + x_{\beta 1}R \subset K_{\beta-1} + x_{\beta 1}R + x_{\beta 2}R \subset \ldots$ and set $K_\beta = K_{\beta-1} + \cup_{k=1}^{\infty} I_{\beta k}$ where $I_{\beta k} = \sum_{i=1}^{k} x_{\beta i}R$.

iv. If β is a limit ordinal, let $K_\beta = \cup_{\alpha < \beta} K_\alpha$.

We will prove that $\oplus_{\alpha \in \Delta} M_\alpha$ is injective by transfinite induction. Let β be the smallest ordinal number such that $K_\beta = K$ and note that if $\beta = 1$, then

$K = t_\tau(R)$. Since $\oplus_{\alpha \in \Delta} M_\alpha \in \mathscr{F}$, it follows that $f = 0$. Therefore, f can obviously be extended to R. Thus, suppose that $\beta > 1$ and make the induction hypothesis that for each ordinal $\zeta < \beta$ every R–linear mapping $K_\zeta \to \oplus_{\alpha \in \Delta} M_\alpha$ can be extended to R. There are three cases to be considered:

Case I: β is not a limit ordinal and $K/K_{\beta - 1}$ is finitely generated. By the induction hypothesis, the mapping $f|_{K_{\beta - 1}}$ can be extended to R. Since this extension evaluated at 1 has at most a finite number of nonzero components in $\oplus_{\alpha \in \Delta} M_\alpha$, this extension is a finite homomorphism. But since $K/K_{\beta - 1}$ is finitely generated, it follows that f is a finite homomorphism. Hence, f is a mapping into an injective summand of $\oplus_{\alpha \in \Delta} M_\alpha$ and so can be extended to R.

Case II: β is not a limit ordinal and $K/K_{\beta - 1}$ is not finitely generated. Then by iii above $K = K_{\beta - 1} + \cup_{k=1}^{\infty} I_{\beta k}$ Since $t_\tau(R) \subseteq K_{\beta - 1}$, we have a strictly increasing chain $\{ J_{\beta k} \}_{k \geq 1}$ of right ideals of R such that $t_\tau(R) \subset J_{\beta 1} \subset J_{\beta 2} \subset J_{\beta 3} \subset \ldots$ where $J_{\beta k} = K_{\beta - 1} + I_{\beta k}$ for each i. By the induction hypothesis, $f | K_{\beta - 1}$ can be extended to R and so it must be the case that $f | K_{\beta - 1}$ is a finite homomorphism. Hence, $f | J_{\beta k}$ is a finite homomorphism for each $k \geq 1$. Next, note that $K = \cup_{k=1}^{\infty} J_{\beta k}$ and by 5 there is an integer $n \geq 1$ such that $\left(\cup_{k=1}^{\infty} J_{\beta k} \right) / J_{\beta n} \in \mathscr{T}$. Consequently, by Proposition 4.1.7, f can be extended to R.

Case III: β is a limit ordinal. Now β is the smallest ordinal such that $K_\beta = K$ and so $K = \cup_{\alpha < \beta} K_\alpha$. Suppose that f is not a finite homomorphism. Then there is a sequence of ordinal numbers $\{ \gamma_i \}_{i \geq 1}$ such that:

1. $\gamma_i < \beta$ for all i,
2. $i < j$ implies $\gamma_i < \gamma_j$, and
3. $f(K_{\gamma_i})$ has nonzero coordinates in at least i of the $\{ M_\alpha \}_{\alpha \in \Delta}$.

If $\cup_{i=1}^{\infty} K_{\gamma_i}$ is properly contained in $K = K_\beta$, then there is an ordinal number $\mu < \beta$ such that $\cup_{i=1}^{\infty} K_{\gamma_i} \subseteq K_\mu$. But then $f|_{K_\mu}$ cannot be a finite homomorphism which contradicts our induction hypothesis that $f|_{K_\mu}$ extends to R which would ensure that $f|_{K_\mu}$ is a finite homomorphism. Hence, we have that $K = \cup_{i=1}^{\infty} K_{\gamma_i}$. We now conclude from 5 and Proposition 4.1.7 that f can be extended to R. □

Proposition 4.1.9 *If* τ *is a cohereditary torsion theory on* Mod–R, *then* $R/t_\tau(R)$ *is a right noetherian ring if and only if every direct sum of* τ–torsionfree *injective* R–*modules is injective.*

Proof. Suppose every direct sum of τ–torsionfree injective R–modules is injective. Let $K_1/t_\tau(R) \subseteq K_2/t_\tau(R) \subseteq K_3/t_\tau(R) \subseteq \ldots$ be an increasing chain of right ideals of $R/t_\tau(R)$. Since τ is cohereditary, $(R/t_\tau(R))/(K_i/t_\tau(R)) \cong R/K_i$ is τ–torsionfree for each i. Thus, $K_1 \subseteq K_2 \subseteq K_3 \subseteq \ldots$ is an ascending chain of τ–pure right ideals of R. Since every direct sum of τ–torsionfree injective R–modules is injective, Proposition 4.1.8 shows that there is an integer n such that $\left(\cup_{i=1}^{\infty} K_i\right)/K_n \in \mathscr{T}$. But $\left(\cup_{i=1}^{\infty} K_i\right)/K_n \in \mathscr{F}$ since K_n is τ–pure in R. Hence, $\left(\cup_{i=1}^{\infty} K_i\right)/K_n = 0$ which implies that the chain $K_1 \subseteq K_2 \subseteq \ldots$ stabilizes. Consequently, the original chain stabilizes and so $R/t_\tau(R)$ is noetherian.

Conversely, suppose that $R/t_\tau(R)$ is noetherian and let $\{ M_\alpha \}_{\alpha \in \Delta}$ be a family of τ–torsionfree injective R–modules. Note that $M_\alpha t_\tau(R) \subseteq t_\tau(M) = 0$ and so M_α is an $R/t_\tau(R)$–module for each $\alpha \in \Delta$. Moreover, it is not difficult to show that each M_α is an injective $R/t_\tau(R)$–module. Hence, by the result of Matlis, $\oplus_{\alpha \in \Delta} M_\alpha$ is an injective $R/t_\tau(R)$–module. The proof will be complete if we can show that $\oplus_{\alpha \in \Delta} M_\alpha$ is an injective R–module. Suppose that K is a right ideal of R and let $f : K \to \oplus_{\alpha \in \Delta} M_\alpha$ be an R–linear mapping, Since $\oplus_{\alpha \in \Delta} M_\alpha \in \mathscr{F}$, $f(t_\tau(K)) = 0$ and so we have an induced mapping

$f^* : K/t_\tau(K) \to \oplus_{\alpha \in \Delta} M_\alpha$. Since $t_\tau(K) = K \cap t_\tau(R)$, we see that $K/t_\tau(K) = K/(K \cap t_\tau(R)) \cong (K + t_\tau(R))/t_\tau(R)$ as R–modules and so we can consider $K/t_\tau(K)$ to be an R–submodule of $R/t_\tau(R)$. It follows from $Kt_\tau(R) \subseteq t_\tau(K)$, that $K/t_\tau(K)$ is a right ideal of the ring $R/t_\tau(R)$. Hence, the mapping f^* can be extended to an $R/t_\tau(R)$–linear mapping $g : R/t_\tau(R) \to \oplus_{\alpha \in \Delta} M_\alpha$ by the $R/t_\tau(R)$–injectivity of $\oplus_{\alpha \in \Delta} M_\alpha$. Since g is also R–linear, if $\pi : R \to R/t_\tau(R)$ is the canonical surjection, $g \circ \pi$ extends f to R. Therefore, $\oplus_{\alpha \in \Delta} M_\alpha$ is an injective R–module. □

Definition 4.1.10 A torsion theory τ on Mod–R is said to be *stable* if $M \in \mathcal{T}$ implies that $E(M) \in \mathcal{T}$.

Proposition 4.1.11 *If* τ *is a stable cohereditary torsion theory on* Mod–R, *then* $t_\tau(M)$ *is a direct summand of* M *for every* R–*module* M.

Proof. If $t_\tau(M)$ is essential in M, then M can be embedded in $E(t_\tau(M))$ and so $M \in \mathcal{T}$ since τ is stable. Thus, $t_\tau(M) = M$ and so $t_\tau(M)$ is trivially a direct summand of M. If $t_\tau(M)$ is not an essential submodule of M, via Zorn's Lemma, choose N to be maximal among the submodules X of M such that $t_\tau(M) \cap X = 0$. Then $N \oplus t_\tau(M)$ is essential in M and so it follows that $(N \oplus t_\tau(M))/N$ is essential in M/N. Now $(N \oplus t_\tau(M))/N \cong t_\tau(M)$ and from this it follows that $(N \oplus t_\tau(M))/N \subseteq t_\tau(M/N)$. Consequently, $t_\tau(M/N)$ is essential in M/N and so $M/N \in \mathcal{T}$. This in turn implies that $M/(N \oplus t_\tau(M)) \in \mathcal{T}$ since $\mathcal{T}$ is closed under homomorphic images. Also note $M/(N \oplus t_\tau(M)) \in \mathcal{F}$ since $M/(N \oplus t_\tau(M))$ is a homomorphic image of $M/t_\tau(M) \in \mathcal{F}$. Thus, $M/(N \oplus t_\tau(M)) \in \mathcal{T} \cap \mathcal{F} = 0$ and so $M = N \oplus t_\tau(M)$. □

Corollary 4.1.12 *If* τ *is a stable cohereditary torsion theory on* Mod–R, *then the following are equivalent:*

1. *Every direct sum of* τ*–torsionfree injective* R*–modules is injective.*
2. $R = t_\tau(R) \oplus S$ *(ring direct sum) where* S *is a right noetherian ring.*

Proof. The proof is immediate by Propositions 4.1.9 and 4.1.11. □

Definition 4.1.13 If τ is a torsion theory on Mod–R, $F_\tau(R)$ is said to have a *cofinal subset of finitely generated right ideals* if for each $K \in F_\tau(R)$ there is a finitely generated right ideal $I \in F_\tau(R)$ such that $I \subseteq K$.

Proposition 4.1.14 *If* τ *is a torsion theory on* Mod–R *such that every direct sum of* τ*–torsionfree injective* R*–modules is injective, then* $F_\tau(R)$ *has a cofinal subset of finitely generated right ideals.*

Proof. Let $K \in F_\tau(R)$. Construct a transfinite strictly increasing sequence of right ideals of R contained in K as follows:

i. If $\beta \geq 1$ is not a limit ordinal and $K/K_{\beta-1}$ is finitely generated, set $K_\beta = K$. ($K_0 = 0$)

ii. $\beta \geq 1$ is not a limit ordinal and K/K_β is not finitely generated. Set $K_\beta = K_{\beta-1} + \cup_{k=1}^{\infty} I_{\beta k}$ where $I_{\beta k} = \sum_{i=1}^{k} x_{\beta i}R$ and the $x_{\beta i} \in K$, $i = 1, 2, 3, \ldots$, are chosen to be such that $I_{\beta 1} \subset I_{\beta 2} \subset I_{\beta 3} \subset \ldots$. ($K_0 = 0$.)

iii. If β is a limit ordinal, set $K_\beta = \cup_{\alpha<\beta} K_\alpha$.

We proceed by transfinite induction. We claim that for each ordinal β there is a finitely generated right ideal J_β of R contained in K such that $K_\beta/J_\beta \in \mathcal{T}$. If $\beta = 1$ and K is finitely generated, set $J_1 = K_1 = K$. If $\beta = 1$, and K is not finitely generated, then $K_1 = \cup_{k=1}^{\infty} I_{1k}$. Since direct sums of

τ–torsionfree injective R–modules are injective, Proposition 4.1.8 provides for an integer $n \geq 1$ such that $K_1/I_{1n} \in \mathcal{T}$. Hence, I_{1n} is a finitely generated right ideal of R contained in K_1 such that $K_1/I_{1n} \in \mathcal{T}$. Set $J_1 = I_{1n}$.

Next, suppose that $\beta > 1$ and make the induction hypothesis that for each ordinal number $\alpha < \beta$ there is a finitely generated right ideal J_α of R contained in K_α such that $K_\alpha/J_\alpha \in \mathcal{T}$. If β is not a limit ordinal there are two cases to be considered:

Case I: $K/K_{\beta-1}$ is finitely generated. Suppose $\{ y_{\beta 1} + K_{\beta-1}, y_{\beta 2} + K_{\beta-1}, y_{\beta 3} + K_{\beta-1}, \ldots, y_{\beta N} + K_{\beta-1} \}$ is a set of generators for $K/K_{\beta-1}$. Set $J_\beta = J_{\beta-1} + \sum_{t=1}^{N} y_{\beta t}R$. Since $J_{\beta-1}$ is finitely generated, it is obvious that J_β is finitely generated. Let $J = \sum_{t=1}^{N} y_{\beta t}R$ and note that $J_{\beta-1} \subseteq K_{\beta-1}$ implies that $K_{\beta-1} \cap (J_{\beta-1} + J) = J_{\beta-1} + (K_{\beta-1} \cap J)$. Now $K_{\beta-1}/J_{\beta-1} \in \mathcal{T}$ and so the R–linear epimorphism

$$K_{\beta-1}/J_{\beta-1} \to K_{\beta-1}/(J_{\beta-1} + (K_{\beta-1} \cap J)) = (K_{\beta-1} + J)/(J_{\beta-1} + J)$$

tells us that $(K_{\beta-1} + J)/(J_{\beta-1} + J) \in \mathcal{T}$. Hence, it follows from the exactness of $0 \to \dfrac{K_{\beta-1} + J}{J_{\beta-1} + J} \to \dfrac{K_\beta}{J_{\beta-1} + J} \to \dfrac{K_\beta}{K_{\beta-1} + J} \to 0$ that $K_\beta/J_\beta \in \mathcal{T}$.

Case II: $K/K_{\beta-1}$ is not finitely generated. In this case, $K_\beta = K_{\beta-1} + \cup_{k=1}^{\infty} I_{\beta k}$. Since $I_{\beta 1} \subset I_{\beta 2} \subset I_{\beta 3} \subset \ldots$ is strictly increasing, by Proposition 4.1.8, there is an integer $n \geq 1$ such that $\left(\cup_{k=1}^{\infty} I_{\beta k} \right)/I_{\beta n} \in \mathcal{T}$. Set $J_\beta = J_{\beta-1} + I_{\beta n}$. Using an argument similar to that in Case I, we see that J_β is a finitely generated right ideal of R contained in K_β such that $K_\beta/J_\beta \in \mathcal{T}$.

Finally, suppose that β is a limit ordinal. Then $K_\beta = \cup_{\alpha < \beta} K_\alpha$. For each ordinal $\alpha < \beta$ choose a right ideal I_α of R containing K_α such that $I_\alpha/K_\alpha = t_\tau(R/K_\alpha)$. Then each I_α is a τ–pure right ideal of R. From the I_α's use induction to choose a sequence as follows: Let γ_1 be the first ordinal $\leq \beta$, if it exists, such that $K_{\gamma_1} \not\subset I_1$, let γ_2 be the first limit ordinal $\leq \beta$ such that $K_{\gamma_2} \not\subset I_{\gamma_1}$, etc. Thus, we obtain a strictly increasing sequence $I_1 \subset I_{\gamma_1} \subset I_{\gamma_2} \subset \cdots$ of τ–pure right ideal of R. But direct sums of τ–torsionfree injective R–modules are injective and so, by Proposition 4.1.8, R is τ–noetherian. Hence, this ascending chain of τ–pure right ideals of R must stabilize. Consequently, there is an integer m such that $I_{\gamma_m} = I_{\gamma_{m+1}} = \cdots$ and so $K_\alpha \subseteq I_{\gamma_m}$ for each ordinal α such that $\gamma_{m+1} < \alpha < \beta$. Hence, $K_\beta \subseteq I_{\gamma_m}$. Since $\mathcal{T}$ is closed under submodules and $K_\beta/K_{\gamma_m} \subset I_{\gamma_m}/K_{\gamma_m} = t_\tau(R/K_{\gamma_m})$, $K_\beta/K_{\gamma_m} \in \mathcal{T}$. By the induction hypothesis, there is finitely generated right ideal J_{γ_m} of R contained in K_{γ_m} such that $K_{\gamma_m}/J_{\gamma_m} \in \mathcal{T}$. It now follows from the exactness of $0 \to K_{\gamma_m}/J_{\gamma_m} \to K_\beta/J_{\gamma_m} \to K_\beta/K_{\gamma_m} \to 0$ that $K_\beta/J_{\gamma_m} \in \mathcal{T}$. Hence, if we let $J_\beta = J_{\gamma_m}$ the case for limit ordinals is finished.

Finally, if β is the smallest ordinal for which $K_\beta = K$, then the exactness of $0 \to K/J_\beta \to R/J_\beta \to R/K \to 0$ shows that $J_\beta \in F_\tau(R)$. □

The Generalized Fuchs Condition

Definition 4.1.15 If M is an R–module, $\Omega(M)$ will denote the set of all right ideals of R which contain the right annihilator of an element of M.

Recall that an R–module is quasi–injective if for any submodule N of M, each $f \in \mathrm{Hom}_R(N, M)$ can be extended to a $g \in \mathrm{Hom}_R(M, M)$. This is obviously equivalent to the requirement that whenever L and N are submodules of M with $L \subseteq N$, any $f \in \mathrm{Hom}_R(L, M)$ can be extended to a $g \in \mathrm{Hom}_R(N, M)$. However, as we will see, these two equivalent formulations of quasi–injective modules lead to different versions of quasi–injectivity relative to a torsion theory.

Fuchs has shown in [18] that quasi–injective modules are characterized by a condition that is similar to Baer's Condition. He has shown that an R–module M is quasi–injective if and only if for each right ideal K of R, each map $f \in \mathrm{Hom}_R(K, M)$ with $\ker f \in \Omega(M)$ can be extended to R. We will refer to this as The Fuchs Condition. Results in [5] show that one way to generalize Fuchs' result on quasi-injective modules is to define quasi–injective modules relative to a torsion theory τ as follows.

Definition 4.1.16 An R–module M is said to be $\overset{f}{\tau}$*–quasi–injective* if, whenever L and N are submodules of M such that $L \in F_\tau(N)$, any $f \in \mathrm{Hom}_R(L, M)$ can be extended to a map $g \in \mathrm{Hom}_R(N, M)$.

We need the following proposition in order to generalize Fuchs' result.

Proposition 4.1.17 *The following are equivalent for an* R*–module* M.

1. M *is* $\overset{f}{\tau}$*–quasi–injective.*
2. *If* $L \subseteq mR \subseteq M$ *and* $L \in F_\tau(mR)$, *then any map* $f \in \mathrm{Hom}_R(L, M)$ *can be extended to a map* $g \in \mathrm{Hom}_R(mR, M)$.
3. *If* L *and* N *are* R*–modules, not necessarily submodules of* M, *such that* $L \in F_\tau(N)$ *and* $\Omega(N) \subseteq \Omega(M)$, *any map* $f \in \mathrm{Hom}_R(L, M)$ *can be extended to a map* $g \in \mathrm{Hom}_R(N, M)$.

Proof. $1 \Rightarrow 2$ and $3 \Rightarrow 1$ are obvious, so we show $2 \Rightarrow 3$. Let L, N and $f \in \mathrm{Hom}_R(L, M)$ be as in 3. Suppose that $\mathcal{S}$ is the set of all pairs (X, g) such that $L \subseteq X \subseteq N$ and $g \in \mathrm{Hom}_R(X, M)$ extends f to X. Partial order $\mathcal{S}$

by $(X, g) \leq (Y, h)$ if and only if $X \subseteq Y$ and h extends g to Y. Via Zorn's Lemma choose (N^*, g^*) to be a maximal element of $\mathcal{S}$. If $N \neq N^*$ and $x \in N \setminus N^*$, then $(0 : x) \in \Omega(N) \subseteq \Omega(M)$, so let $m \in M$ be such that $(0 : m) \subseteq (0 : x)$. This gives an R–map $h : m(N^* : x) \to M : mr \to g^*(xr)$. Since $N^* \in F_\tau(N)$ implies that $(N^* : x) \in F_\tau(R)$, it follows that $m(N^* : x) \in F_\tau(mR)$. Consequently, h extends linearly to a map $k : mR \to M$ which in turn yields an R–linear mapping $\psi : N^* + xR \to M : y + xr \to g^*(y) + k(mr)$. Therefore, $(N^*, g^*) < (N^* + xR, \psi)$ which is a contradiction. Hence, $N = N^*$ and so we have the implication. □

Proposition 4.1.18 (**The Generalized Fuchs Condition**) *An* R*–module* M *is* $\overset{f}{\tau}$*–quasi–injective if and only if for every right ideal* $K \in F_\tau(R)$ *each* $f \in \mathrm{Hom}_R(K, M)$ *with* $\ker f \in \Omega(M)$ *can be extended to* R.

Proof. Suppose M is $\overset{f}{\tau}$–quasi–injective and let $f \in \mathrm{Hom}_R(K, M)$ where $K \in F_\tau(R)$ and $\ker f \in \Omega(M)$. Let $m \in M$ be such that $(0 : m) \subseteq \ker f$ and consider $h : mK \to M : mr \to f(r)$. Since $K \in F_\tau(R)$, $mK \in F_\tau(mR)$ and so, by 2 of Proposition 4.1.17, h can be extended linearly to $k : mR \to M$. Hence, if $g : R \to M$ is given by $g(r) = k(mr)$, then g extends f.

Conversely, let $L \in F_\tau(N)$ where N is a submodule of M. If $f \in \mathrm{Hom}_R(L, M)$, let $\mathcal{S}$ be as in the proof of the proposition above. Choose (N^*, g^*) to be maximal in $\mathcal{S}$. If $N \neq N^*$, let $x \in N \setminus N^*$. Then $(N^* : x) \in F_\tau(R)$ since $N^* \in F_\tau(N)$. Now consider the R–linear mapping $h : (N^* : x) \to M : r \to g^*(xr)$. Since $(0 : x) \subseteq \ker h$, $\ker h \in \Omega(M)$. Thus, by hypothesis, h can be extended to $k : R \to M$. This in turn yields $\psi : N^* + xR \to M : y + xr \to g^*(y) + k(r)$. But then $(N^*, g^*) < (N^* + xR, \psi)$

which contradicts the maximality of (N^*, g^*). Hence, $N = N^*$ which shows that M is $\overset{f}{\tau}$–quasi–injective. □

Next, we define τ–quasi–injective modules.

Definition 4.1.19 An R–module M is said to be *τ–quasi–injective*, if whenever $N \in F_\tau(M)$, any $f \in Hom_R(N, M)$ can be extended to a map $g \in Hom_R(M, M)$.

Obviously, any $\overset{f}{\tau}$–quasi–injective R–module is τ–quasi–injective. When τ is the torsion theory in which every module is torsion, both concepts reduce to that of a quasi–injective module.

It follows from the definition of a τ–quasi–injective module that an R–module M is τ–quasi–injective if and only if $Ext^1_R(M/N, M) = 0$ for all $N \in F_\tau(M)$. This in turn is equivalent to $Ext^1_R(M/E, M) = 0$ for all τ–essential submodules E of M.

The following proposition yields additional information regarding τ–completely reducible rings.

Proposition 4.1.20 The following are equivalent for any ring R.

1. R *is τ–completely reducible.*
2. *Every* R*–module is* $\overset{f}{\tau}$*–quasi–injective.*
3. *Every* R*–module is τ–quasi–injective.*

Proof. 1 ⇒ 2. This follows from Proposition 4.1.5 and the fact that any τ–injective R–module is $\overset{f}{\tau}$–quasi–injective.

$2 \Rightarrow 3$. Obvious.

$3 \Rightarrow 1$. In light of Proposition 4.1.5, it suffices to show that 3 implies that every R–module is τ–injective. For any R–module M and $K \in F_\tau(R)$, it follows that $M \times K \in F_\tau(M \times R)$. If every R–module is τ–quasi–injective, then $M \times R$ is a τ–quasi–injective. Consequently, if $f \in \mathrm{Hom}_R(K, M)$, then the R–linear mapping $g : M \times K \to M \times R : (m, k) \to (f(k), 0)$ can be extended to an R–linear mapping $h : M \times R \to M \times R$. But then $h|_R$ extends f to R. Hence, by The Generalized Baer Condition, M is τ–injective. □

4.2 (Quasi–)Projective Modules Relative to a Torsion Theory

We now investigate τ–projective [6] and τ–quasi–projective [7] modules. τ–projective and τ–quasi–projective modules reduce to projective and quasi–projective modules, respectively, when τ is chosen to be the torsion theory in which every module is torsionfree.

Definition 4.2.1 M is said to be τ–*projective* if every row exact diagram of R–modules and R–module homomorphisms of the form

$$\begin{array}{c} M \\ \downarrow \\ 0 \to L \to X \to N \to 0 \end{array}$$

with $L \in \mathscr{F}$ can be completed commutatively by an R–linear map $M \to X$. An R–linear epimorphism $f : M \to N$ is said to be τ–*minimal* if ker f is small and $\ker f \in \mathscr{F}$.

If every R–module is torsionfree, the τ–minimal epimorphisms are just the minimal epimorphisms of Bass [3]. Standard arguments can be used to show that a direct sum of τ–projective R–modules is τ–projective if and only if each summand is τ–projective. Note that an R–module M is τ–projective if and only if $\mathrm{Ext}^1_R(M, N) = 0$ for all $N \in \mathscr{F}$.

Proposition 4.2.2 *The following are equivalent:*

1. $R/t_\tau(R)$ *is a semisimple (artinian) ring.*
2. *Every* R–*module in* $\mathscr{F}$ *is injective.*
3. *Every* R–*module is* τ–*projective.*
4. *Every cyclic* R–*module is* τ–*projective.*

Proof. $1 \Rightarrow 2$. Suppose $M \in \mathscr{F}$. Since $Mt_\tau(R) \subseteq t_\tau(M) = 0$, M is an injective $R/t_\tau(R)$–module. If K is a right ideal of R, let $f \in \mathrm{Hom}_R(K, M)$. Notice that if $x \in t_\tau(K) = K \cap t_\tau(R)$, then $f(x) = 0$ since $f(t_\tau(K)) \subseteq t_\tau(M) = 0$. Thus, we have a map $g : (K + t_\tau(R))/t_\tau(R) \to M$ defined by $g(x + t_\tau(R)) = f(x)$ which is easily seen to be $R/t_\tau(R)$–linear. Consequently, by Baer's Condition there is an $m \in M$ such that $g(x + t_\tau(R) = m(x + t_\tau(R))$ for all $x + t_\tau(R)$ in $R/t_\tau(R)$. But this gives $f(x) = mx$ for all $x \in K$ and so M is injective.

$2 \Rightarrow 3$. Consider the diagram

$$\begin{array}{ccccc} & & M & & \\ & & \downarrow g & & \\ L & \xrightarrow{f} & N & \to & 0 \end{array}$$

where $K = \ker f \in \mathcal{F}$. Since K is injective, $L = K \oplus K'$ for some submodule K' of L. But then $h = f|_{K'} : K' \to N$ is an isomorphism and $\varphi = h^{-1} \circ g$ completes the diagram commutatively.

$3 \Rightarrow 4$ is obvious.

$4 \Rightarrow 1$. Let $K/t_\tau(R)$ be a right ideal of $R/t_\tau(R)$. Since $K/t_\tau(R) \in \mathcal{F}$ and R/K is a τ–projective R–module, the exact sequence of R–modules $0 \to K/t_\tau(R) \to R/t_\tau(R) \to R/K \to 0$ splits. Hence, $K/t_\tau(R)$ is an R–direct summand of $R/t_\tau(R)$ and consequently, an $R/t_\tau(R)$–summand of $R/t_\tau(R)$. Thus, $R/t_\tau(R)$ is semisimple artinian. □

Corollary 4.2.3 *If τ is faithful, every R–module is τ–projective if and only if R is a semisimple (artinian) ring.*

It has been shown in [31] that if $f \in \mathrm{Hom}_R(M, N)$ and X is a small submodule of M, then $f(X)$ is a small submodule of N. Indeed, suppose X is a small submodule of M and let Y be a submodule of N such that $f(X) + Y = N$. Then $X + f^{-1}(Y) = M$ and so $f^{-1}(Y) = M$. Thus, $f(X) \subseteq Y$ and so $Y = N$. This fact is required in proof of the following proposition.

Proposition 4.2.4 *If $f : M \to N$ is a τ–minimal epimorphism, then f is an* isomorphism whenever N is τ–projective.

Proof. Since $K = \ker f \in \mathcal{F}$, the exact sequence $0 \to K \to M \to N \to 0$ splits whenever N is τ–projective. Hence, $M = K \oplus M'$ where M' is a submodule of M which is isomorphic to N. But K is small, so $K = 0$. □

Proposition 4.2.5 *Let* $K = \ker f$ *where* $f : M \to N$ *is an epimorphism and* M *is* τ*–projective. Then* N *is* τ*–projective when* $K \in \mathscr{T}$*. Conversely, if* N *is* τ*–projective and* K *is small in* M*, then* $K \in \mathscr{T}$*.*

Proof. Consider the row exact diagram

$$\begin{array}{ccccc} & & M & & \\ & & \downarrow f & & \\ & & N & & \\ & & \downarrow g & & \\ L & \xrightarrow{h} & X & \to & 0 \end{array}$$

where $\ker h \in \mathscr{F}$. Since M is τ–projective $g \circ f$ factors through L. Suppose $\varphi : M \to L$ is such that $h \circ \varphi = g \circ f$. Now $\varphi(K) \subseteq \ker h$ and so if $K \in \mathscr{T}$, $\varphi(K) \subseteq t_\tau(\ker h) = 0$. Hence, we have an induced map $\varphi^*: N \to L$ which completes the diagram commutatively and so N is τ–projective.

Conversely, suppose N is τ–projective and K is small in M. Then $\psi : M/t_\tau(K) \to N : x + t_\tau(K) \to f(x)$ is an epimorphism and $\ker \psi = K/t_\tau(K) \in \mathscr{F}$. Moreover, by our observations immediately preceding Proposition 4.2.4, $\ker \psi$ is small in $M/t_\tau(K)$ and so ψ is τ–minimal. Hence, ψ must be an isomorphism by Proposition 4.2.4. Consequently, $K = t_\tau(K)$. □

Proposition 4.2.6 M *is* τ*–projective if and only if* M *is isomorphic to a direct summand of* F/K *where* F *is a free* R*–module on* M *and* $K \in \mathscr{T}$*.*

Proof. Let M be τ–projective and suppose that $0 \longrightarrow K \to F \xrightarrow{f} M \to 0$ is a free module on M and $K = \ker f$. Consider the exact sequence $0 \to K/t_\tau(K) \to F/t_\tau(K) \to M \to 0$. Since M is τ–projective and $K/t_\tau(K) \in \mathcal{F}$ this sequence splits.

Conversely, if $0 \to K \to F \to M \to 0$ is a free module on M and $K \in \mathcal{T}$, then F/K is τ–projective because of Proposition 4.2.5. Hence, if M is isomorphic to a direct summand of F/K, then M is τ–projective since a direct summand of a τ–projective module is τ–projective. □

Definition 4.2.7 We say that *the category* Mod–R *has enough* τ*–projective modules* if for every R–module N, there is a τ–projective R–module M and an epimorphism $M \xrightarrow{f} N$ such that $\ker f \in \mathcal{F}$.

Proposition 4.2.8 *The category* Mod–R *has enough* τ*–projective modules.*

Proof. Let $g : F \to N$ be a free module on N and suppose that $K = \ker g$. Since $K/t_\tau(K) \in \mathcal{F}$ and since $F/t_\tau(K)$ is, by Proposition 4.2.5, τ–projective, the proposition follows from the fact that we have an induced epimorphism $\varphi : F/t_\tau(K) \to N$ with $\ker \varphi = K/t_\tau(K) \in \mathcal{F}$. □

Proposition 4.2.9 *Let* M *be such that* $Mt_\tau(R) = 0$. *Then* M *is projective as an* $R/t_\tau(R)$*–module if and only if* M *is* τ*–projective as an* R*–module.*

Proof. Let $\oplus_{\alpha \in I} (R/t_\tau(R))_\alpha \to M \to 0$ be a free $R/t_\tau(R)$–module on M. If M is projective as an $R/t_\tau(R)$–module, then $\oplus_{\alpha \in I} (R/t_\tau(R))_\alpha \cong M \oplus N$ for some $R/t_\tau(R)$–module N. Now $\oplus_{\alpha \in I} (R/t_\tau(R))_\alpha \cong (\oplus_{\alpha \in I} R_\alpha) /(\oplus_{\alpha \in I} t_\tau(R)_\alpha)$ and it follows from Proposition 4.2.5 that $(\oplus_{\alpha \in I} R_\alpha) /(\oplus_{\alpha \in I} t_\tau(R)_\alpha)$

is a τ–projective R–module since $\oplus_{\alpha \in I} t_\tau(R)_\alpha \in \mathcal{T}$. Thus, M is τ–projective as an R–module since a direct summand of a τ–projective R–module is τ–projective.

Conversely, suppose that M is a τ–projective R–module. If $0 \to K \to F \to M \to 0$ is a free $R/t_\tau(R)$–module on M, then $K \in \mathcal{F}$. Consequently, since M is τ–projective, $F \cong M \oplus N$ for some R–module N isomorphic to K. Hence, M is a projective $R/t_\tau(R)$–module. □

Proposition 4.2.10 M *is a* τ*–projective* R*–module if and only if* $M/Mt_\tau(R)$ *is a projective* $R/t_\tau(R)$*–module.*

Proof. If M is a τ–projective R–module, then since $Mt_\tau(R) \in \mathcal{T}$, $M/Mt_\tau(R)$ is, by Proposition 4.2.5, τ–projective as an R–module. Thus, by Proposition 4.2.9, $M/Mt_\tau(R)$ is a projective $R/t_\tau(R)$–module.

Conversely, let $M/Mt_\tau(R)$ be projective as an $R/t_\tau(R)$–module. By Proposition 4.2.8, Mod–R has enough τ–projective modules and so there is a short exact sequence $0 \to K \xrightarrow{i} X \xrightarrow{p} M \to 0$ with $K \in \mathcal{F}$ and X τ–projective. Thus, we have a commutative diagram

$$\begin{array}{ccccccccc}
0 & \to & K & \xrightarrow{i} & X & \xrightarrow{p} & M & \to & 0 \\
 & & \downarrow \lambda'' & & \downarrow \lambda' & & \downarrow \lambda & & \\
0 & \to & (K + Xt_\tau(R))/Xt_\tau(R) & \xrightarrow{j} & X/Xt_\tau(R) & \xrightarrow{q} & M/Mt_\tau(R) & \to & 0
\end{array}$$

where the rows are exact and λ'', λ' and λ are the canonical maps and j and q are the maps induced by i and p, respectively. Since $M/Mt_\tau(R)$ is a projective $R/t_\tau(R)$–module, the bottom row splits. Let $q'\colon M/Mt_\tau(R) \to X/Xt_\tau(R)$

be such that $q \circ q' = 1$. If $f = q' \circ \lambda$, then $q \circ f = q \circ q' \circ \lambda = \lambda$. Suppose $m \in M$. Then $m = p(x)$ for some $x \in X$. Now $q \circ \lambda'(x) = \lambda \circ p(x) = \lambda(m)$ $= q \circ f(m)$ and so $\lambda'(x) - f(m) \in \ker q$. Hence, there exists a $k \in K$ such that $\lambda'(x) - f(m) = j \circ \lambda''(k) = \lambda' \circ i(k)$. If $y = x - i(k)$, then since $K \cap Xt_\tau(R) = 0$, y is the only the only element to satisfy $\lambda'(y) = f(m)$ and $p(y) = m$. If we define $p' : M \to X$ by $p'(m) = y$, then $p \circ p' = 1_M$. Thus, the top row of the diagram splits and so M is τ–projective. □

Corollary 4.2.11 *If* $M = Mt_\tau(R)$, *then* M *is* τ*–projective.*

Definition 4.2.12 A module M is said to be τ–*quasi–projective* if every row exact diagram of R–modules and R–linear mappings of the form

$$\begin{array}{ccccc} & & M & & \\ & & \downarrow & & \\ M & \xrightarrow{f} & N & \to & 0 \end{array}$$

with $\ker f \in \mathscr{F}$ can be completed commutatively by an R–linear mapping $M \to M$.

Proposition 4.2.13 *Let* $\varphi : M \to N$ *be an epimorphism where* M *is* τ*–quasi-projective and* $K = \ker \varphi \in \mathscr{F}$. *If* K *is stable under endomorphisms of* M, *then* N *is* τ*–quasi-projective.*

Proof. Consider the row exact diagram

$$\begin{array}{ccccccc}
 & & & & M & & \\
 & & & & \downarrow \varphi & & \\
 & & h \swarrow & & N & & \\
 & & & h^* \swarrow & \downarrow & & \\
M & \xrightarrow{\varphi} & N & \xrightarrow{f} & X & \rightarrow & 0
\end{array}$$

where $\ker f \in \mathscr{F}$ and suppose that K is stable under endomorphisms of M. If $H = \ker f \circ \varphi$, then $f \circ \varphi(t_\tau(H)) = 0$ and so $\varphi(t_\tau(H)) \subseteq \ker f$. Hence, $\varphi(t_\tau(H)) = 0$ Consequently, $t_\tau(H) \subseteq K$ and so $t_\tau(H) = 0$. Hence, $H \in \mathscr{F}$. Thus, the outer diagram can be completed commutatively by an R–linear mapping h. But $h(K) \subseteq K$ and so we have an induced R–linear mapping h^* which makes the inner diagram commute. □

4.3 Flat Modules Relative to a Torsion Theory

If $\mathscr{C}$ is a class of short exact sequences of modules in Mod–R, then a left R–module M is often said to be flat relative to $\mathscr{C}$ if whenever $0 \rightarrow L \rightarrow X \rightarrow N \rightarrow 0$ is in $\mathscr{C}$, $0 \rightarrow L \otimes_R M \rightarrow X \otimes_R M \rightarrow N \otimes_R M \rightarrow 0$ is exact. If τ is a torsion theory on Mod–R, we consider the class $\mathscr{C}$ of short exact sequences $0 \rightarrow L \rightarrow X \rightarrow N \rightarrow 0$ in Mod–R where $N \in \mathscr{T}$ and investigate left R–modules which are flat relative to $\mathscr{C}$.

Definition 4.3.1 A left R–module M is said to be τ–*flat* if whenever $0 \rightarrow L \rightarrow X \rightarrow N \rightarrow 0$ is a short exact sequence of modules in Mod–R with $N \in \mathscr{T}$, the sequence $0 \rightarrow L \otimes_R M \rightarrow X \otimes_R M \rightarrow N \otimes_R M \rightarrow 0$ is exact.

τ-flat left R-modules [8, 34] become flat left R-modules when τ is the torsion theory in which every R-module is torsion. Consequently, many of the following results become classical results when τ is chosen to be this torsion theory on Mod-R

If $0 \to L \to X \to N \to 0$ is exact with $N \in \mathcal{T}$, the sequence $L \otimes_R M \to X \otimes_R M \to N \otimes_R M \to 0$ is always exact. Thus, to show M is a τ-flat left R-module, it suffices to show that $0 \to L \otimes_R M \to X \otimes_R M$ is exact. Consequently, a left R-module M is τ-flat if and only if $\mathrm{Tor}_1^R(N, M) = 0$ for all $N \in \mathcal{T}$.

It is well known that a left R-module M if flat if and only if its *character module* $M^+ = \mathrm{Hom}_Z(M, Z/Q)$ is an injective R-module where M^+ is made into an R-module by defining $(fr)(m) = f(rm)$. A similar relation exists between τ-flat left R-modules and τ-injective R-modules.

Proposition 4.3.2 *A left* R*-module is* τ*-flat if and only if its character module* M^+ *is a* τ*-injective* R*-module.*

Proof. Suppose that M is a τ-flat left R-module and consider the short exact sequence $0 \to L \to X \to N \to 0$ where $N \in \mathcal{T}$. Since the sequence $0 \to L \otimes_R M \to X \otimes_R M \to N \otimes_R M \to 0$ is exact, it follows from the fact that Q/Z is an injective Z-module that the sequence $\mathrm{Hom}_Z(X \otimes_R M, Q/Z) \to \mathrm{Hom}_Z(L \otimes_R M, Q/Z) \to 0$ is exact. This leads to the commutative row exact diagram

$$\begin{array}{ccc} \mathrm{Hom}_Z(X \otimes_R M, Q/Z) & \to & \mathrm{Hom}_Z(L \otimes_R M, Q/Z) \to 0 \\ \alpha_X \downarrow & & \alpha_L \downarrow \\ \mathrm{Hom}_R(X, \mathrm{Hom}_Z(M, Q/Z)) & \to & \mathrm{Hom}_R(L, \mathrm{Hom}_Z(M, Q/Z)) \to 0 \end{array}$$

where $[\alpha_X(f)(x)](m) = f(x \otimes m)$ and $[\alpha_L(f)(l)](m) = f(l \otimes m)$ are the canonical isomorphisms. Hence, $\mathrm{Hom}_R(X, M^+) \to \mathrm{Hom}_R(L, M^+) \to 0$ is exact and so it follows that M^+ is τ–injective. Since this argument is easily reversible, the converse holds. □

We have seen (Proposition 4.1.2) that the (essential) right ideals of $F_\tau(R)$ can be used as a test set for the τ–injectivity of an R–module. The following proposition shows that the (essential) right ideals of $F_\tau(R)$ can also be used as a test set for the τ–flatness of a left R–module.

Proposition 4.3.3 *The following are equivalent:*

1. M *is a* τ*–flat left* R*–module.*
2. *If* $K \in F_\tau(R)$ *and* $0 \to K \to R$ *is the canonical injection, the sequence* $0 \to K \otimes_R M \to R \otimes_R M$ *is exact.*
3. *If* E *is a* τ*–essential right ideal of* R *and* $0 \to E \to R$ *is the canonical injection, the sequence* $0 \to E \otimes_R M \to R \otimes_R M$ *is exact.*

Proof. $1 \Rightarrow 2$ and $2 \Rightarrow 3$ are obvious, so suppose E is a τ–essential right ideal of R. Since $R/E \in \mathcal{T}$, it follows that the sequence

$0 \to E \otimes_R M \to R \otimes_R M \to R/E \otimes_R M \to 0$ is exact. By using an argument similar to that in the proof of Proposition 4.3.2, we can conclude that the sequence $0 \to \mathrm{Hom}_R(R/E, M^+) \to \mathrm{Hom}_R(R, M^+) \to \mathrm{Hom}_R(E, M^+) \to 0$ is exact. Consequently, every R–linear mapping from a τ–essential right ideal of R to M^+ can be extended to R and so, by Corollary 4.1.3, M^+ is τ–injective. Therefore, Proposition 4.3.2 shows that M is τ–flat and so $3 \Rightarrow 1$. □

Corollary 4.3.4 *For a left R–module* M, $\mathrm{Tor}_1^R(R/K, M) = 0$ *for all (essential) right ideals* $K \in F_\tau(R)$ *if and only if* M *is* τ*–flat.*

If K is a right ideal of R and M is a left R–module, φ_K will denote the Z–linear epimorphism $K \otimes_R M \to KM : \sum_{i=1}^n k_i \otimes m_i \to \sum_{i=1}^n k_i m_i$. If there is a need to emphasize the module M, the mapping will be denoted by φ_K^M.

The following two propositions establish additional connections between τ–flat left R–modules and the right ideals in $F_\tau(R)$.

Proposition 4.3.5 *The following are equivalent:*

1. M *is a* τ*–flat left* R*–module.*
2. *For each* $K \in F_\tau(R)$, φ_K *is an isomorphism.*
3. *For each* τ*–essential right ideal* E *of* R, φ_E *is an isomorphism.*

Proof. $1 \Rightarrow 3$. If $E \in F_\tau(R)$ is essential and $j : E \to R$ is the canonical injection, consider the commutative diagram

$$\begin{array}{ccc} E \otimes_R M & \xrightarrow{j \otimes 1_M} & R \otimes_R M \\ \varphi_E \downarrow & & \varphi_R \downarrow \\ EM & \xrightarrow{j^\#} & RM = M \end{array}$$

where $j^\#$ is the obvious mapping. Since φ_R is an isomorphism, $j \otimes 1_M = \varphi_R{}^{-1} \circ j^\# \circ \varphi_E$. Thus, $\varphi_E\left(\sum_{i=1}^n e_i \otimes m_i\right) = 0$ leads to $j \otimes 1_M\left(\sum_{i=1}^n e_i \otimes m_i\right) = 0$ which in turn gives $\sum_{i=1}^n e_i \otimes m_i = 0$ since $j \otimes 1_M$ is a monomorphism. Hence, the epimorphism φ_E is an isomorphism.

$3 \Rightarrow 2$. Let $K \in F_\tau(R)$ and choose K' to be maximal among all the right ideals J of R such that $K \cap J = 0$. Then $K \oplus K' \in F_\tau(R)$ and $K \oplus K'$ is essential in R. Hence, the canonical map $\varphi_{K \oplus K'} : (K \oplus K') \otimes_R M \to (K \oplus K')M$ is an isomorphism. But $(K \otimes_R M) \oplus (K' \otimes_R M) \cong (K \oplus K') \otimes_R M$ and so if $K \otimes_R M$ is identified with its image in $(K \oplus K') \otimes_R M$, $\varphi_K = \varphi_{K \oplus K'}|_{K \otimes_R M}$. Consequently, φ_K is an isomorphism.

$2 \Rightarrow 1$. Replace E by K in the diagram of $1 \Rightarrow 3$. If φ_K is an isomorphism, $j \otimes_R 1_M$ is a monomorphism since $j^{\#}$ is a monomorphism and φ_R^{-1} is an isomorphism. □

Proposition 4.3.6 *If* M *is a left* R*–module and* $\pi : F \to M$ *is a free left* R*–module on* M *with* $N = \ker \pi$, *the following are equivalent:*

1. M *is* τ*–flat.*
2. $KF \cap N = KN$ *for each* $K \in \mathcal{F}_\tau(R)$.
3. $EF \cap N = EN$ *for each* τ*–essential right ideal* E *of* R.

Proof. $1 \Rightarrow 2$. If $K \in \mathcal{F}_\tau(R)$, $K \otimes_R N \overset{1_K \otimes j}{\to} K \otimes_R F \overset{1_K \otimes \pi}{\to} K \otimes_R M \to 0$ is exact where $j : N \to F$ is the canonical injection. Consider the row exact commutative diagram

$$\begin{array}{ccccccc} & & K \otimes_R N & \overset{1_K \otimes j}{\to} & K \otimes_R F & \overset{1_K \otimes \pi}{\to} & K \otimes_R M \to 0 \\ & & & & \varphi_K^F \downarrow & & \varphi_K^M \downarrow \\ 0 \to & KF \cap N & & \to & KF & \overset{\theta}{\to} & KM \to 0 \end{array}$$

where $\theta : KF \to KM : \sum_{i=1}^n k_i f_i \to \sum_{i=1}^n k_i \pi(f_i)$. Since F is free, F is a τ–flat left R–module so, by Proposition 4.3.5, φ_K^F is an isomorphism. Notice also that θ is an epimorphism, since π is an epimorphism. Since $KF \subseteq F$, $\theta = \pi|_{KF}$ and so $\ker \theta = KF \cap N$. Also, by Proposition 4.3.5, φ_K^M is an isomorphism, since we are assuming that M is a τ–flat left R–module. Let rf be a generator of $\ker\theta = KF \cap N$. Then $\varphi_K^M \circ (1_K \otimes \pi) \circ \varphi_K^{F^{-1}}(rf) = 0$ and so $(1_K \otimes \pi) \circ \varphi_K^{F^{-1}}(rf) = 0$ because φ_K^M is an isomorphism. From this it follows that $\varphi_K^{F^{-1}}(rf) \in \ker(1_K \otimes \pi) = \text{Image}(1_K \otimes j)$. Thus, there is an element $\sum_{i=1}^n k_i \otimes g_i \in K \otimes_R N$ such that $(1_K \otimes j)\left(\sum_{i=1}^n k_i \otimes g_i\right) = \sum_{i=1}^n k_i \otimes g_i = \varphi_K^{F^{-1}}(rf)$. Hence, $\varphi_K^F\left(\sum_{i=1}^n k_i \otimes g_i\right) = rf$ and so $rf = \sum_{i=1}^n k_i g_i \in KN$. Therefore, $KF \cap N \subseteq KN$ and so $KF \cap N = KN$.

$2 \Rightarrow 3$ is obvious and so the proof will be finished if we can show $3 \Rightarrow 1$. Consider the commutative diagram of $1 \Rightarrow 2$ with K replaced by a τ–essential right ideal E of R. Suppose $EF \cap N = EN$. Since φ_E^M is an epimorphism, we are required only to show that φ_E^M is a monomorphism. Let $e \otimes m$ be a generator of $\ker \varphi_E^M$. Since π is an epimorphism, there is an $f \in F$ such that $\pi(f) = m$. Hence, $\varphi_E^M \circ (1_E \otimes \pi)(e \otimes f) = 0$ and so $\theta \circ \varphi_E^F(e \otimes f) = 0$. Thus, $\theta(ef) = 0$ which implies that $ef \in EF \cap N = EN$. If $ef = \sum_{i=1}^n e_i g_i$, where $e_i \in E$ and $g_i \in N$ for each i, $\varphi_E^{F-1}(ef) = \varphi_E^{F-1}\left(\sum_{i=1}^n e_i g_i\right)$ and so $e \otimes f = \sum_{i=1}^n e_i \otimes g_i$. Therefore, $e \otimes m = (1_E \otimes \pi)(e \otimes_R f) = (1_E \otimes \pi)\left(\sum_{i=1}^n e_i \otimes g_i\right) = \sum_{i=1}^n e_i \otimes \pi(g_i) = 0$. Consequently, φ_E^M is a monomorphism and so, by Proposition 4.3.5, M is τ–flat. □

If R is an integral domain, the canonical torsion theory τ on Mod–R is defined as follows: an element m of an R–module M is torsion if there exists a nonzero element $r \in R$ such that $mr = 0$. If $t_\tau(M)$ is the set of all τ–torsion elements of M, then $t_\tau(M)$ is a submodule of M. Moreover, the classes $\mathscr{T} = \{ M \mid t_\tau(M) = M \}$ and $\mathscr{F} = \{ M \mid t_\tau(M) = 0 \}$ define this torsion theory on Mod–R which is hereditary and faithful. Over this torsion theory on Mod–R it is well known that if R is a principal ideal domain, an R–module is flat if and only if it is in $\mathscr{F}$. We need the following proposition in order to prove a modest generalization of this observation.

Proposition 4.3.7 *Let* R *be a commutative ring and* τ *a torsion theory on* Mod–R. *If the ideals of* $F_\tau(R)$ *are principal, then every module in* $\mathscr{F}$ *is* τ*–flat.*

Proof. Suppose $M \in \mathscr{F}$ and let $\pi : F \to M$ be a free R–module on M with $\ker \pi = N$. By Proposition 4.3.6, to show that M is τ–flat, it suffices to show $kF \cap N = kN$ for each $k \in R$ for which $(k) \in F_\tau(R)$. If k is such an element of R and $kf \in kF \cap N$, then $f + N \in F/N \cong M$ is such that $(k)(f + N) = 0$ and so $f + N \in t_\tau(F/N) = 0$. Thus, $f \in N$ and so $kf \in kN$. Hence, $kF \cap N \subseteq kN$ and so equality follows. □

Proposition 4.3.8 *Let* R *be a commutative ring and* τ *a torsion theory on* Mod–R. *If the ideals of* $F_\tau(R)$ *are principal, then* $\mathscr{F}$ *coincides with the class of* τ*–flat modules if and only if* τ *is faithful.*

Proof. Suppose that the class $\mathscr{F}$ coincides with the class of τ–flat modules. Since R is a projective module, R is flat and consequently τ–flat. Thus, $R \in \mathscr{F}$ and so τ is faithful.

Conversely, suppose that τ is faithful. In view of Proposition 4.3.7, we are only required to show that every τ–flat module is in $\mathscr{F}$. Let $\pi : F \to M$

be a free module on M with ker π = N and suppose that M is a τ–flat R–module. If $f + N \in t_\tau(F/N)$, there is a principal right ideal $(k) \in F_\tau(R)$ such that $(k)(f + N) = 0$. Consequently, $kf \in N$ and so $kf \in kF \cap N = kN$. If $kf = kg$ where $g \in N$, then $(k)(f - g) = 0$. Thus, $f - g \in t_\tau(F)$. Now by Proposition 1.1.9, $\mathscr{F}$ is closed under submodules and direct products and so since $R \in \mathscr{F}$, it follows that every free R–module is in $\mathscr{F}$. Thus, $f = g$ and so $f + N = 0$. Hence, $t_\tau(M) \cong t_\tau(F/N) = 0$. □

Proposition 4.3.9 *If* τ *is such that* $\tau_G \leq \tau$, *where* τ_G *is the Goldie torsion theory, the following are equivalent:*

1. M *is a flat left* R–*module.*
2. M *is a* τ–*flat left* R–*module.*
3. M *is a* τ_G–*flat left* R–*module.*

Proof. (See 1.1.20, Example 3, for details on the Goldie torsion theory.) $1 \Rightarrow 2$ is obvious and $2 \Rightarrow 3$ follows immediately from $F_G(R) \subseteq F_\tau(R)$ and Proposition 4.3.6. For $3 \Rightarrow 1$, notice that in Proposition 4.3.6 if we select τ to be the torsion theory in which every module is torsion, a left R–module M is flat if and it is τ–flat. Consequently, a left R–module is flat if and only if $EF \cap N = EN$ for each essential right ideal of R where $\pi : F \to M$ is a free left R–module on M with ker π = N. But since $F_G(R)$ contains all the essential right ideals of R, $3 \Rightarrow 1$, again by Proposition 4.3.6. □

Corollary 4.3.10 *If* R *is a commutative principal ideal ring and* τ_G *is faithful, then the class of flat* R–*modules coincides with the class* $\mathscr{F}_G$.

Proof. 1. The proposition above shows that an R–module is flat if and only if it is τ_G –flat and it follows from Proposition 4.3.8 that the class of τ_G–flat R–modules coincides with the class $\mathscr{F}_G$. □

Since projective modules are always flat, it seems reasonable to suspect that τ–flat left R–modules might somehow be connected to left R–modules which are projective relative to some torsion theory on R–Mod. The following shows that this is indeed the case.

Proposition 4.3.11 *If* τ *is a torsion theory on* Mod–R, *then the collection* $\mathscr{D} = \{ M \in R\text{–Mod} \mid N \otimes_R M = 0 \ \ \forall N \in \mathscr{T} \}$ *is a torsion class in* R–Mod.

Proof. By 1 of Proposition 1.9, $\mathscr{D}$ will be a torsion class in R–Mod provided $\mathscr{D}$ is closed under homomorphic images, extensions and direct sums. $\mathscr{D}$ is closed under direct sums since if $\{ M_{\alpha \in I} \}$ is a family of leftR–modules each of which is in $\mathscr{D}$ and $N \in \mathscr{T}$, then $N \otimes_R (\bigoplus_{\alpha \in I} M_\alpha) \cong \bigoplus_{\alpha \in I} (N \otimes_R M_\alpha) = 0$. The proof that $\mathscr{D}$ is closed under homomorphic images and extensions is equally straightforward. □

Definition 4.3.12 Let τ be a torsion theory on Mod–R and suppose that $\mathscr{D} = \{ M \in R\text{–Mod} \mid N \otimes_R M = 0 \ \ \forall M \in \mathscr{T} \}$. If $\mathscr{R}$ is the torsionfree class corresponding to $\mathscr{D}$, the torsion theory $(\mathscr{D}, \mathscr{R})$ on R–Mod will be denoted by $\tau_\otimes$. We will often refer to $\tau_\otimes$ as the torsion theory on R–Mod which corresponds to the torsion theory τ on Mod–R.

In 1.1.20, Example 5, we have seen that $\tau_\otimes$ is the torsion theory $(\mathscr{D}, \mathscr{R})$ of τ–divisible and τ–reduced left R–modules. *Observe that* $\tau_\otimes$ *may not be hereditary even though (we are assuming that)* τ *is hereditary.* If every R–module in $\mathscr{T}$ is a flat, then of course $\tau_\otimes$ will be hereditary. The following proposition appears in [34].

Proposition 4.3.13 *If* τ *is a torsion theory on* Mod–R *with corresponding torsion theory* $\tau_\otimes$ *on* R–Mod, *then the following hold:*

1. *If* M *is a* τ*–flat left* R*–module and* $A \subseteq M$ *with* $A \in \mathscr{D}$, *then* M/A *is* τ*–flat.*
2. *If* M *is a* $\tau_\otimes$*–projective left* R*–module, then* M *is* τ*–flat.*
3. *If* $M/t_{\tau_\otimes}(R)M$ *is a flat left* $R/t_{\tau_\otimes}(R)$*–module, then* $M/t_{\tau_\otimes}(R)M$ *is a* τ*–flat left* R*–module.*

Proof. 1. The short exact sequence $0 \to A \to M \to M/A \to 0$ gives rise to an exact sequence $\mathrm{Tor}_1^R(N, M) \to \mathrm{Tor}_1^R(N, M/A) \to N \otimes_R A$ for any $N \in \mathscr{T}$. But $\mathrm{Tor}_1^R(N, M) = 0$ for all $N \in \mathscr{T}$ since M is τ–flat and $N \otimes_R A = 0$ because $A \in \mathscr{D}$. Thus, $\mathrm{Tor}_1^R(N, M/A) = 0$ for all $N \in \mathscr{T}$ and so M/A is τ–flat.

2. Consider the short exact sequence $0 \to K \to P \to M \to 0$ where P is a projective left R–module. This yields a short exact sequence $0 \to K/t_{\tau_\otimes}(K) \to P/t_{\tau_\otimes}(K) \to M \to 0$ where $K/t_{\tau_\otimes}(K) \in \mathscr{R}$. Since M is a $\tau_\otimes$–projective left R–module, this sequence splits. But by 1, $P/t_{\tau_\otimes}(K)$ is τ–flat because $t_{\tau_\otimes}(K) \in \mathscr{D}$. Since a direct summand of a τ–flat left R–module is τ–flat, M is τ–flat.

3. It is not difficult to show that R–Mod has enough $\tau_\otimes$–projectives even though $\tau_\otimes$ may not be hereditary (The proofs of Propositions 4.2.5 and 4.2.8 do not depend on τ being hereditary.) and so we can find a short exact sequence $0 \to K \to P \to M/t_{\tau_\otimes}(R)M \to 0$ with P $\tau_\otimes$–projective and $K \in \mathscr{R}$. First, observe that $t_{\tau_\otimes}(R)P \in \mathscr{D}$. This follows from the fact that the canonical epimorphism $N \otimes_R t_{\tau_\otimes}(R) \to Nt_{\tau_\otimes}(R)$ gives $Nt_{\tau_\otimes}(R) = 0$ when $N \in \mathscr{T}$. Thus, if $N \in \mathscr{T}$ and $n \otimes rp$ is a generator of $N \otimes_R t_{\tau_\otimes}(R)P$, $n \otimes rp = nr \otimes p = 0 \otimes p = 0$. Hence, $t_{\tau_\otimes}(R)P \in \mathscr{D}$. Consequently, $t_{\tau_\otimes}(R)P \subseteq K$ implies that $t_{\tau_\otimes}(R)P = 0$ and so $0 \to K \to P \to M/t_{\tau_\otimes}(R)M \to 0$ is an exact sequence of left $R/t_{\tau_\otimes}(R)$–modules. For any $N \in \mathscr{T}$, we have a commutative diagram

$$\begin{array}{ccccccc} 0 = \mathrm{Tor}_1^R(N, P) & \to & \mathrm{Tor}_1^R(N, M/t_{\tau_\otimes}(R)M) & \to & N \otimes_R K & \to & N \otimes_R P \\ & & & & \downarrow & & \downarrow \\ & & 0 = \mathrm{Tor}_1^{R/\tau_\otimes(R)}(N, M/t_{\tau_\otimes}(R)M) & \to & N \otimes_{R/t_{\tau_\otimes}(R)} K & \to & N \otimes_{R/t_{\tau_\otimes}(R)} P \end{array}$$

where the vertical arrows are isomorphisms. The top row is exact since P is a $\tau_\otimes$–projective left R–module which is, by 2, τ–flat . The bottom row is exact since we are assuming that $M/t_{\tau_\otimes}(R)M$ is a flat left $R/t_{\tau_\otimes}(R)$–module. Thus, $\mathrm{Tor}_1^R(N, M/t_{\tau_\otimes}(R)M) = 0$ for all $N \in \mathscr{T}$ and so $M/t_{\tau_\otimes}(R)M$ is a τ–flat left R–module. □

Corollary 4.3.14 *When $\tau_\otimes$ is hereditary, a left R–module M is τ–flat if and only if $M/t_{\tau_\otimes}(R)M$ is a τ–flat left R–module.*

Proof. If M is a τ–flat left R–module, then, by 1 of Proposition 4.3.13, $t_{\tau_\otimes}(R)M \in \mathscr{D}$ implies that $M/t_{\tau_\otimes}(R)M$ is a τ–flat left R–module. Conversely, assume that $M/t_{\tau_\otimes}(R)M$ is a τ–flat left R–module and let $0 \to K \to P \to M \to 0$ be a short exact sequence with P $\tau_\otimes$–projective and $K \in \mathscr{R}$. This yields an exact sequence $0 \to (K + t_{\tau_\otimes}(R)P)/t_{\tau_\otimes}(R)P \to P/t_{\tau_\otimes}(R)P \to M/t_{\tau_\otimes}(R)M \to 0$. But since $\tau_\otimes$ is hereditary, $K \cap t_{\tau_\otimes}(R)P \in \mathscr{D} \cap \mathscr{R} = 0$ and so $(K + t_{\tau_\otimes}(R)P)/t_{\tau_\otimes}(R)P \cong K/(K \cap t_{\tau_\otimes}(R)P) \cong K$. Thus, for any $N \in \mathscr{T}$ we have a commutative diagram

$$\begin{array}{ccccccc} 0 = \mathrm{Tor}_1^R(N, P) \to & \mathrm{Tor}_1^R(N, M) & \to & N \otimes_R K & \to & N \otimes_R P \\ & & & \| & & \downarrow \\ & 0 = \mathrm{Tor}_1^R(N, M/t_{\tau_\otimes}(R)M) & \to & N \otimes_R K & \to & N \otimes_R P/t_{\tau_\otimes}(R)P \end{array}$$

where the top row is exact by virtue of 2 of Proposition 4.3.13. The bottom row is exact since $M/t_{\tau_\otimes}(R)M$ is a τ–flat left R–module. Thus, $\mathrm{Tor}_1^R(N, M) = 0$ for all $N \in \mathscr{T}$ and so M is a τ–flat left R–module. □

Corollary 4.3.15 *If* $\tau_\otimes$ *is hereditary and* $M/t_{\tau_\otimes}(R)M$ *is flat as a left* $R/t_{\tau_\otimes}(R)$*–module, then* M *is a* τ*–flat left* R*–module.*

Proof. If $M/t_{\tau_\otimes}(R)M$ is a flat left $R/t_{\tau_\otimes}(R)$–module, then, by Proposition 4.3.13, $M/t_{\tau_\otimes}(R)M$ is a τ–flat left R–module. Hence, by Corollary 4.3.14, M is a τ–flat left R–module. □

Bass [3] has shown that a ring R is left perfect if and only if the class of flat left R–modules coincides with the class of projective left R–modules. The converse of the following proposition will be given after τ–projective covers have been developed.

Proposition 4.3.16 *Let* τ *is a torsion theory such that* $\tau_{\otimes}$ *is hereditary. If the class of* $\tau_{\otimes}$*–projective left* R*–modules coincides with the class of* τ*–flat left* R*–modules, then* $R/t_{\tau_{\otimes}}(R)$ *is left perfect.*

Proof. Let M be a flat left $R/t_{\tau_{\otimes}}(R)$–module. Then, by Corollary 4.3.15, M is a τ–flat left R–module. But then, by our assumption, M is a $\tau_{\otimes}$–projective left R–module. Therefore, by Proposition 4.2.10, M is a projective $R/t_{\tau_{\otimes}}(R)$–module. Hence, $R/t_{\tau_{\otimes}}(R)$ is a left perfect ring. □

§5 COVERS AND HULLS

In this section we study various covers and hulls of modules in the setting of a torsion theory on Mod–R. We show that the injective hull of Eckmann and Schopf [11] and the quasi–injective hull of Johnson and Wong [26] can be reproduced in the setting of a general torsion theory on Mod–R.

As pointed out in §1, a ring R is said to be *right perfect* if every R–module has a *projective cover*. Bass [3] has characterized right perfect rights as being precisely those rings for which $R/J(R)$ is semisimple and $J(R)$ is *right T–nilpotent*. A subset I of R is right T–nilpotent, if for every sequence $x_1, x_2, \cdots$ in I, $x_n \cdots x_2x_1 = 0$ for some integer $n \geq 1$. For a torsion theory τ on Mod–R, Rangaswamy has shown [33] that the τ–projective covers [6] can be used to determine when $R/t_\tau(R)$ is right perfect. He showed that $R/t_\tau(R)$ is right perfect if and only if every R–module has a τ–projective cover. Fuller and Hill [17] and Koehler [28] have independently shown that right perfect rings can be characterized by using the quasi–projective covers of Wu and Jans [38]. They showed that a ring R is right perfect if and only if every R–module has a quasi–projective cover. This result can also be generalized in the setting of a general torsion theory on Mod–R. It was shown in [7] that $R/t_\tau(R)$ is right perfect if and only if every R–module has a τ–quasi–projective cover. As we shall see these results reduce to the result of Bass and to the result of Fuller–Hill and Koehler when τ is chosen to be the torsion theory in which every R–module is torsionfree.

We will also investigate torsionfree covers. Torsionfree covers were first investigated by Enochs [12] when τ is the usual torsion theory on Mod–R associated with an integral domain. In this setting torsionfree covers were shown to universally exist. These covers were later generalized [35] in the setting of an arbitrary torsion theory on Mod–R.

5.1 (Quasi–)Injective Hulls Relative to a Torsion Theory

In §1 we pointed out that the injective hull [11] of an R–module M is an injective R–module E(M) together with an injective R–linear mapping $\varphi : M \to E(M)$ such that $\varphi(M)$ is an essential submodule of M. It is well known that every R–module has an injective hull which is unique up to isomorphism. Furthermore, the injective hull of M is minimal among the injective extensions of M. It is common practice to identify M with its image under φ so that $M \subseteq E(M)$. No loss of generality results from this identification. We will now show that, given a torsion theory τ on Mod–R, every R–modules has a τ–injective hull which is also unique up to an isomorphism.

Definition 5.1.1 A τ–injective R–module N together with an injective R–linear mapping $f : M \to N$ is said to be a τ–*injective hull* of M if f(M) is a τ–essential submodule of N. A τ–injective hull of M will be denoted by $\varphi : M \to E_\tau(M)$ whenever it can be shown to exist.

Proposition 5.1.2 *Every* R–*module* M *has a* τ–*injective hull which is unique up to isomorphism.*

Proof. Let $E_\tau(M) = \{ m \in E(M) \mid (M : m) \in F_\tau(R) \}$ where E(M) denotes the injective hull of M. Then $E_\tau(M)$ is a submodule of E(M) which contains M as an essential submodule. Since $(M : m) = (0 : m + M)$, we see that

$(0 : m + M) \in F_\tau(R)$ whenever $m + M \in E_\tau(M)/M$. Thus, $E_\tau(M)/M \in \mathcal{T}$ and so $M \in F_\tau(E_\tau(M))$. Thus, M is τ–essential in $E_\tau(M)$. Next, we claim $E(M)/E_\tau(M) \in \mathcal{F}$. If $m + E_\tau(M)$ is a τ–torsion element of $E(M)/E_\tau(M)$, then $(E_\tau(M) : m) = (0 : m + E_\tau(M)) \in F_\tau(R)$. Thus, if $r \in (E_\tau(M) : m)$, then $mr \in E_\tau(M)$ indicates that $((M : m) : r) = (M : mr) \in F_\tau(R)$. It now follows from 2 of Definition 1.1.12 that $(M : m) \in F_\tau(R)$. Hence, $m \in E_\tau(M)$ and so $m + E_\tau(M) = 0$. Now consider the row exact commutative diagram

$$\begin{array}{ccccccccc} 0 & \to & L & \to & N & \to & N/L & \to & 0 \\ & & f \downarrow & & g \downarrow & & h \downarrow & & \\ 0 & \to & E_\tau(M) & \to & E(M) & \to & E(M)/E_\tau(M) & \to & 0 \end{array}$$

where $L \in F_\tau(N)$, f is a given R–linear mapping, g is the R–linear mapping whose existence is ensured by the injectivity of $E(M)$ and h is the map defined by $x + L \to g(x) + E_\tau(M)$. Since $N/L \in \mathcal{T}$ and $E(M)/E_\tau(M) \in \mathcal{F}$, $h = 0$. Hence, $g(N) \subseteq E_\tau(M)$ and so $E_\tau(M)$ is τ–injective. Consequently, $\varphi : M \to E_\tau(M)$ is a τ–injective hull of M where φ is the canonical injection.

Finally, suppose $\varphi : M \to E_\tau(M)$ and $\varphi^* : M \to E_\tau(M)^*$ are arbitrary τ–injective hulls of M. Then we have a commutative diagram

$$\begin{array}{ccc} & & E_\tau(M) \\ & \varphi \nearrow & \downarrow \psi \\ M & \overset{\varphi^*}{\to} & E_\tau(M)^* \end{array}$$

where ψ is given by the τ–injectivity of $E_\tau(M)^*$. If $\varphi(m) \in \varphi(M) \cap \ker \psi$, then $\varphi^*(m) = \psi \circ \varphi(m) = 0$ and so $m = 0$. Thus, $\varphi(M) \cap \ker \psi = 0$ and so

$\ker \psi = 0$ since $\varphi(M)$ is essential in $E_\tau(M)$. Consequently, ψ is an injective mapping. Next, consider the short exact sequence
$0 \to \psi(E_\tau(M)) \to E_\tau(M)^* \to E_\tau(M)^*/\psi(E_\tau(M)) \to 0$. Since $\psi(E_\tau(M))$ is τ–injective and $\psi(E_\tau(M)) \in F_\tau(E_\tau(M)^*)$, this sequence splits. If $E_\tau(M)^* = \psi(E_\tau(M)) \oplus X$, then $\varphi^*(M) \cap X = 0$. Hence, $X = 0$ since $\varphi^*(M)$ is an essential submodule of $E_\tau(M)^*$. Therefore, ψ is an isomorphism. □

$E_\tau(M)$ is the smallest τ–injective R–module containing M. This follows from the fact that if $M \subseteq N \subseteq E_\tau(M)$ where N is τ–injective, then the short exact sequence $0 \to N \xrightarrow{j} E_\tau(M) \to E_\tau(M)/N \to 0$ splits. Hence, if $j : N \to E_\tau(M)$ is the canonical injection, $E_\tau(M) = N \oplus \ker f$ where $f : E_\tau(M) \to N$ is such that $f \circ j = 1_M$. Since N is essential in $E_\tau(M)$, $\ker f = 0$ and so $N = E_\tau(M)$. Since M is τ–essential in $E_\tau(M)$ it also follows that $E_\tau(M)$ embeds in very τ–injective extension of M.

If σ and τ are torsion theories on Mod–R such that $\sigma \le \tau$, there is an R–linear injection $\psi : E_\sigma(M) \to E_\tau(M)$ such that the diagram

$$\begin{array}{ccc} & & E_\sigma(M) \\ & \varphi_\sigma \nearrow & \downarrow \psi \\ M & \overset{\varphi_\tau}{\to} & E_\tau(M) \end{array}$$

is commutative. Thus, the isomorphism classes of relative injective hulls of M can be partially ordered by using the partial ordering on the torsion theories on Mod–R.

If $\varphi : M \to E_\tau(M)$ is an arbitrary τ–injective hull of M, we identify M with $\varphi(M)$ and consider M to be a submodule of $E_\tau(M)$. In this situation it is often the case that mapping φ will not be specifically mentioned and $E_\tau(M)$

will be referred to as "the" τ–injective hull of M. As in the case for injective hulls, this identification results in no loss of generality. The following proposition describes additional connections between $E_\sigma(M)$ and $E_\tau(M)$ when σ and τ are torsion theories on Mod–R such that $\sigma \leq \tau$.

Proposition 5.1.3 *If* $E_\sigma(M)$ *and* $E_\tau(M)$ *are relative injective hulls of* M *where* σ *and* τ *are torsion theories on* Mod–R *such that* $\sigma \leq \tau$, *then:*

1. $E_\sigma(M)$ *is the largest submodule of* $E_\tau(M)$ *containing* M *such that* $E_\sigma(M)/M = t_\sigma(E_\tau(M)/M)$. *That is,* $E_\sigma(M)$ *is the* τ*–pure closure of* M *in* $E_\tau(M)$.
2. $E_\sigma(M) = \{ x \in E_\tau(M) \mid (M : x) \in F_\sigma(R) \}$
3. *If* σ *is cogenerated by the injective module* E, *then* $E_\sigma(M) = \bigcap_{f \in \Delta} \ker f$ *where* $\Delta = \{ f : E_\tau(M) \to E \mid f(M) = 0 \}$.

Proof. 1. Let $E_\sigma(M)$ be the largest submodule of $E_\tau(M)$ containing M such that $E_\sigma(M)/M = t_\sigma(E_\tau(M)/M)$. Then M is σ–essential in $E_\sigma(M)$ and so we need only show that $E_\sigma(M)$ is σ–injective. First, note that $E_\tau(M)/E_\sigma(M) \cong (E_\tau(M)/M)/(E_\sigma(M)/M) = (E_\tau(M)/M)/(t_\sigma(E_\tau(M)/M)) \in \mathcal{F}_\sigma$. Hence, if $K \in F_\sigma(R)$ and $f : K \to E_\sigma(M)$ is R–linear, we have a commutative diagram

$$\begin{array}{ccccccccc} 0 & \to & K & \to & R & \to & R/K & \to & 0 \\ & & \downarrow f & & \downarrow g & & \downarrow h & & \\ 0 & \to & E_\sigma(M) & \to & E_\tau(M) & \to & E_\tau(M)/E_\sigma(M) & \to & 0 \end{array}$$

where g is the mapping given by the τ–injectivity of $E_\tau(M)$ and h is the induced mapping. Since $R/K \in \mathcal{T}_\sigma$ and $E_\tau(M)/E_\sigma(M) \in \mathcal{F}_\sigma$, $h = 0$ and so it follows that $g(R) \subseteq E_\sigma(M)$.

2. $t_\sigma(E_\tau(M)/M) = \{ x+M \in E_\tau(M)/M \mid (0 : x+M) \in F_\sigma(R) \}$

$$= \{ x+M \in E_\tau(M)/M \mid (M : x) \in F_\sigma(R) \}$$

$$= \{ x \in E_\tau(M) \mid (M : x) \in F_\sigma(R) \}/M.$$

Hence, $E_\sigma(M) = \{ x \in E_\tau(M) \mid (M : x) \in F_\sigma(R) \}$.

3. If $x \in E_\sigma(M)$, then by 2, $(M : x) \in F_\sigma(R)$. Hence, if $f \in \Delta$, $f(x)(M : x) = f(x(M : x)) = 0$ since $x(M : x) \subseteq M$ and $f(M) = 0$. Thus, $f(x) \in t_\sigma(E) = 0$ and so $x \in \cap_{f \in \Delta} \ker f$.

Conversely, suppose that $x \in \cap_{f \in \Delta} \ker f$ where $x \in E_\tau(M)$. By 2 to show that $x \in E_\sigma(M)$, it suffices to show that $(M : x) \in F_\sigma(R)$. But $(M : x)$ will be in $F_\sigma(R)$ provided $\mathrm{Hom}_R(R/(M : x), E) = 0$. Let $g \in \mathrm{Hom}_R(R/(M : x), E)$ and consider the commutative diagram

$$\begin{array}{ccccc} & & & & E_\tau(M) \\ & & & & \downarrow \eta \\ 0 & \rightarrow & R/(M : x) & \xrightarrow{k} & E_\tau(M)/M \\ & & \downarrow g & \swarrow h & \\ & & E & & \end{array}$$

where $k(r + (M : x)) = xr + M$, η is the natural surjection and h is the R–linear mapping given by the injectivity of E. Since $h \circ \eta(M) = 0$, $h \circ \eta \in \Delta$ and so $h \circ \eta(x) = 0$. Hence, $g(1 + (M : x)) = h \circ k(1 + (M : x)) = h(x + M) = 0$ and so $g = 0$. Therefore, $x \in E_\sigma(M)$. □

From the above we see that if σ is the torsion theory on Mod–R which is cogenerated by the injective hull $E(M)$ of M, then $E_\sigma(M)$ is the maximal rational extension of M [14, 15, 16]. Indeed, if τ is the torsion theory in which every R–module is torsion, then $\sigma \leq \tau$ and $E_\tau(M) = E(M)$. Hence, 3 of the proposition above produces $E_\sigma(M) = \cap_{f \in \Delta} \ker f$ where $\Delta = \{ f : E(M) \to E(M) \mid f(M) = 0 \}$. But this is, by definition, the maximal rational extension of M.

We previously mentioned that Matlis [30] has shown that R is a right noetherian ring if and only if direct sums of injective R–modules are injective. His result can be recovered from Propositions 4.1.8 and 4.1.9 when τ is chosen to be the torsion theory in which every module is torsionfree. The result of Matlis can also be recovered from the following two propositions when τ is chosen to be the torsion theory in which every module is torsion.

Proposition 5.1.4 *The following are equivalent for a torsion theory* τ *on* Mod–R.

1. *The right ideals in* $F_\tau(R)$ *satisfy the ascending chain condition.*
2. *Direct sums of* τ*–torsion* τ*–injective* R*–modules are* τ*–injective.*
3. *Countable direct sums of* τ*–torsion* τ*–injective* R*–modules are* τ*–injective.*

Proof. $1 \Rightarrow 2$. Let $\{ M_\alpha \}_{\alpha \in \Delta}$ be a family of τ–injective R–modules each of which is in $\mathscr{T}$. If $K \in F_\tau(R)$ and $f : K \to \oplus_{\alpha \in \Delta} M_\alpha$ is R–linear, to show that $\oplus_{\alpha \in \Delta} M_\alpha$ is τ–injective it suffices to show that f is a finite homomorphism. That is, it suffices to show that there is a finite subset B of Δ such that $f(K) \subseteq \oplus_{\alpha \in B} M_\alpha$. For then f is a mapping into a τ–injective summand of $\oplus_{\alpha \in \Delta} M_\alpha$ and so can be extended to R. Since $\mathscr{T}$ is closed under direct sums, $\oplus_{\alpha \in \Delta} M_\alpha \in \mathscr{T}$. Hence, $K/\ker f \in \mathscr{T}$ since $\mathscr{T}$ is closed under submodules. The short exact sequence $0 \to K/\ker f \to R/\ker f \to R/K \to 0$ shows that $R/\ker f \in \mathscr{T}$ since $\mathscr{T}$ is closed under extensions. Hence, $\ker f \in F_\tau(R)$.

Now let's show that f is a finite homomorphism. Suppose $B = \{ \alpha \in \Delta \mid \pi_\alpha \circ f(K) \neq 0 \}$ where $\pi_\beta : \oplus_{\alpha \in \Delta} M_\alpha \to M_\beta$ is the canonical projection. If B is not finite, let $C = \{ \alpha_1, \alpha_2, \dots \}$ be a countably infinite subset of B and for each integer $k \geq 1$, set $C_k = \{ \alpha_1, \alpha_2, \dots, \alpha_k \}$. Then under the appropriate identifications $(\oplus_{\alpha \in (\Delta \setminus C)} M_\alpha) \oplus M_{\alpha_1} \subseteq (\oplus_{\alpha \in (\Delta \setminus C)} M_\alpha) \oplus M_{\alpha_1} \oplus M_{\alpha_2} \subseteq \cdots \subseteq (\oplus_{\alpha \in (\Delta \setminus C) \cup C_k} M_\alpha) \subseteq \cdots \subseteq \oplus_{\alpha \in \Delta} M_\alpha$. Hence, $\{ I_k \}_{k \geq 1}$, where $I_k = f^{-1}(\oplus_{\alpha \in (\Delta \setminus C) \cup C_k} M_\alpha)$, is a strictly increasing chain of right ideals of R each of which is contained in K. Moreover, $K = \cup_{k=1}^{\infty} I_k$. Since $\ker f \subseteq I_k$ for each k, $I_k \in F_\tau(R)$ for $k = 1, 2, \cdots$. But the right ideals in $F_\tau(R)$ satisfy the ascending chain condition and so there is an integer n such that $I_n = I_{n+1} = \cdots$, a contradiction. Thus, B must be finite and $f(K) \subseteq \oplus_{\alpha \in B} M_\alpha$.

$2 \Rightarrow 3$ is obvious.

$3 \Rightarrow 1$. Let $K_1 \subseteq K_2 \subseteq K_3 \subseteq \cdots$ be an ascending chain of right ideals of R each of which is in $F_\tau(R)$. Since $R/K_i \in \mathcal{T}$, it follows that $E_\tau(R/K_i) \in \mathcal{T}$. Hence, $\oplus_{i=1}^{\infty} E_\tau(R/K_i)$ is τ–injective. Next, define $f : K \to \oplus_{i=1}^{\infty} E_\tau(R/K_i)$, where $K = \cup_{i=1}^{\infty} K_i$, by $f(x) = (x + K_i)$. Now $K \in F_\tau(R)$ and so f can be extended to R. If g extends f to R, then $g(1)$ has at most a finite number of nonzero coordinates and $f(x) = g(1)x$ for all $x \in K$. Hence, we see that $(x + K_i) = g(1)x$ and so if the nth coordinate of $g(1)$ is zero, we see that $x + K_n = 0$ for all $x \in K$. From this it follows that $K_n = K_{n+1} = \cdots$ and so the result follows. □

The τ–torsion condition can be removed from the proposition above provided one addition assumption is added to the ascending chain condition on right ideals of $F_\tau(R)$.

Proposition 5.1.5 *The following are equivalent for a torsion theory* τ *on* Mod–R.

1. *The following two conditions hold:*

   A. *The right ideals of* $F_\tau(R)$ *satisfy the ascending chain condition.*

   B. *If* $K_1 \subseteq K_2 \subseteq K_3 \subseteq \cdots$ *is an ascending chain of right ideals of* R *such that* $\cup_{i=1}^{\infty} K_i \in F_\tau(R)$, *then* $K_n \in F_\tau(R)$ *for some integer* $n \geq 1$.

2. *Direct sums of* τ*–injective R–modules are* τ*–injective.*
3. *Countable direct sums of* τ*–injective* R*–modules are* τ*–injective.*

Proof. $1 \Rightarrow 2$. Let $\{ M_\alpha \}_{\alpha \in \Delta}$ be a family of τ–injective R–modules and suppose that $f : K \rightarrow \oplus_{\alpha \in \Delta} M_\alpha$ is R–linear where $K \in F_\tau(R)$. If $f(K) \not\subset \oplus_{\alpha \in B} M_\alpha$ for some finite subset B of Δ, then, as in the proof of Proposition 5.1.4, we can construct a strictly increasing chain $I_1 \subset I_2 \subset I_3 \subset \cdots$ of right ideals of R such that $K = \cup_{i=1}^{\infty} I_i$. Hence, $\cup_{i=1}^{\infty} I_i \in F_\tau(R)$ and so by B, $K_n \in F_\tau(R)$ for some integer $n \geq 1$. Therefore, by A, $I_m = I_{m+1} = \cdots$ for some integer m such that $m \geq n$. But this is a contradiction and so there must be a finite subset B of Δ such that $f(K) \subset \oplus_{\alpha \in B} M_\alpha$. Since $\oplus_{\alpha \in B} M_\alpha$ is τ–injective, this is sufficient to show that $\oplus_{\alpha \in \Delta} M_\alpha$ is τ–injective.

$2 \Rightarrow 3$. Obvious.

$3 \Rightarrow 1$. Let $K_1 \subseteq K_2 \subseteq K_3 \subseteq \cdots$ an ascending chain of right ideals of R such that $K = \cup_{i=1}^{\infty} K_i \in F_\tau(R)$. If $f : K \rightarrow \oplus_{i=1}^{\infty} E_\tau(R/K_i) : x \rightarrow (x + K_i)$, then, by The Generalized Baer Condition, there is an $m \in \oplus_{i=1}^{\infty} E_\tau(R/K_i)$ such that $(x + K_i) = mx$ for all $x \in K$. If the nth coordinate of m is zero, then $x + K_n = 0$ for all $x \in K$ and so $K = K_n$. Hence, $K_n \in F_\tau(R)$ and so we have condition B. Condition A follows immediately by choosing the K_i to be in $F_\tau(R)$. □

We now develop τ–quasi–injective hulls which are reminiscent of the quasi–injective hulls of Johnson and Wong [26].

Definition 5.1.6 A τ–quasi–injective R–module N together with an injective R–linear mapping $f : M \to N$ is said to be a *τ–quasi–injective hull* of M if f(M) is a τ–essential submodule of N. A τ–quasi–injective hull of M will be denoted by $\varphi : M \to Q_\tau(M)$ whenever it can be shown to exist.

The following proposition provides the framework for establishing the existence of τ–quasi–injective hulls.

Proposition 5.1.7 *For any* R*–module* M, *let* $\Delta = \mathrm{End}_R(\, E_\tau(M)\,)$. *Then:*

1. $\Delta M = \left\{ \sum_{i=1}^{n} f_i(m_i) \mid f_i \in \Delta,\ m_i \in M \textit{ for } i = 1, 2, \cdots, n,\ n \textit{ not fixed.} \right\}$ *is the intersection of all the τ–quasi–injective submodules of* $E_\tau(M)$ *containing* M.
2. ΔM *is a τ–quasi–injective* R*–module.*
3. M *is a τ–quasi–injective* R*–module if and only if* $\Delta M = M$.

Proof. 3. Suppose $\Delta M = M$, $N \in F_\tau(M)$ and $f \in \mathrm{Hom}_R(\, N, M\,)$. Then f can be extended to an R–linear mapping $g : M \to E_\tau(M)$ which in turn can be extended to an $h \in \Delta$. But $\Delta M = M$ indicates that $h(M) \subseteq M$ and so $h|_M : M \to M$ extends f. Hence, M is a τ–quasi–injective module.

Conversely, suppose M is τ–quasi–injective R–module and let $f \in \Delta$. Then $M/(\, M \cap f^{-1}(M)\,) \cong (\, M + f^{-1}(M)\,)/M \subseteq E_\tau(M)/M \in \mathscr{T}$ and so $N = M \cap f^{-1}(M) \in F_\tau(M)$. Since $f|_N : N \to M$, by the τ–quasi–injectivity of M, there is an R–linear mapping $g : M \to M$ which extends $f|_N$. In turn there is an $h \in \Delta$ which extends g. If $(\, h - f)(M) \neq 0$, then

$(h - f)(M) \cap M \neq 0$ since M is essential in $E_\tau(M)$. Let $x, y \in M$ be nonzero and such that $(h - f)(x) = y$. Then $(h - f)(x) = (g - f)(x)$ and so $f(x) = g(x) - y \in M$. Therefore, $x \in N$ and so $(g - f)(x) = 0$ which is a contradiction. Thus, $(h - f)(M) = 0$ and therefore $f(M) = h(M) = g(M) \subseteq M$. Hence, $\Delta M \subseteq M$.

2. Since $M \subseteq \Delta M \subseteq E_\tau(M)$, M is a τ–essential submodule of ΔM. Consequently, $E_\tau(M) = E_\tau(\Delta M)$ and so $\Delta = \mathrm{End}_R(E_\tau(\Delta M))$. From this it follows that $\Delta(\Delta M) = \Delta M$. 2 now follows from 3.

1. Let $\mathscr{C}$ be the collection of all τ–quasi–injective submodules of $E_\tau(M)$ which contain M. Since ΔM is a τ–quasi–injective R–module, $\Delta M \in \mathscr{C}$ and so $\cap_{N \in \mathscr{C}} N \subseteq \Delta M$. For the reverse containment, note that $M \subseteq \cap_{N \in \mathscr{C}} N$ implies $\Delta M \subseteq \Delta\left(\cap_{N \in \mathscr{C}} N\right)$ so let $f(n)$ be a generator of $\Delta\left(\cap_{N \in \mathscr{C}} N\right)$ where $f \in \Delta$ and $n \in \cap_{N \in \mathscr{C}} N$. Then $f(n) \in \Delta N$ for all $N \in \mathscr{C}$. Now for any $N \in \mathscr{C}$, $M \subseteq N \subseteq E_\tau(M)$ indicates that M is a τ–essential submodule of N. Therefore, $E_\tau(M) = E_\tau(N)$ and so $\Delta = \mathrm{End}_R(E_\tau(M)) = \mathrm{End}_R(E_\tau(N))$. But N is a τ–quasi–injective R–module and so $\Delta N = N$. Hence, $f(n) \in N$ for each $N \in \mathscr{C}$ and so we have shown that $\Delta\left(\cap_{N \in \mathscr{C}} N\right) \subseteq \cap_{N \in \mathscr{C}} N$. Consequently, $\Delta M \subseteq \cap_{N \in \mathscr{C}} N$. □

Proposition 5.1.8 *Every* R*–module has a* τ*–quasi–injective hull which is unique up to isomorphism.*

Proof. Let ΔM be as in Proposition 5.1.7 and set $Q_\tau(M) = \Delta M$. Since $M \subseteq Q_\tau(M) \subseteq E_\tau(M)$, M is a τ–essential submodule of $Q_\tau(M)$. Moreover, we have seen in Proposition 5.1.7 that $Q_\tau(M)$ is a τ–quasi–injective R–module. Hence, $\varphi : M \to Q_\tau(M)$, where φ is the canonical injection, is a τ–quasi–injective hull of M.

Next we claim that if $f : M \to X$ is a τ–quasi–injective τ–dense extension of M, there is an injective R–linear mapping $\psi : Q_\tau(M) \to X$ such that $\psi \circ \varphi = f$. Suppose $f : M \to X$ is such an extension of M. It follows that there is an R–linear injective mapping $g : E_\tau(M) \to E_\tau(X)$ such that $f(M) \subseteq g(Q_\tau(M)) \subseteq g(E_\tau(M)) \subseteq E_\tau(X)$. Since $E = g(E_\tau(M))$ is a τ–injective hull of the τ–quasi–injective R–module $Q = g(Q_\tau(M))$, if $\bar{\Delta} = \mathrm{Hom}_R(E, E)$, then $\bar{\Delta} Q \subseteq Q$. Now $f(M) \in F_\tau(E_\tau(X))$ and $X \in F_\tau(E_\tau(X))$ implies that $f(M) \in F_\tau(X)$ and so $E \in F_\tau(E_\tau(X))$ since $E \supseteq f(M)$. Consequently, any map $h \in \bar{\Delta}$ can be extended to a map $\hat{h} \in \hat{\Delta}$ where $\hat{\Delta} = \mathrm{Hom}_R(E_\tau(X), E_\tau(X))$. But $\hat{\Delta} X \subseteq X$ and so $h(X) = \hat{h}(X) \subseteq X$. Therefore, $\bar{\Delta} X \subseteq X$. It now follows that $\bar{\Delta}(Q \cap X) \subseteq Q \cap X$ and since $f(M) \subseteq Q \cap X \subseteq E$, E is a τ–injective hull of $Q \cap X$ and so $Q \cap X$ is τ–quasi–injective. Hence, $M \subseteq g^{-1}(Q \cap X) \subseteq E_\tau(M)$ and $g^{-1}(Q \cap X)$ is τ–quasi–injective. We also see that $g^{-1}(Q \cap X) \subseteq g^{-1}(Q) = Q_\tau(M)$ and so $g^{-1}(Q \cap X) = Q_\tau(M)$ since $Q_\tau(M)$ is the smallest τ–quasi–injective submodule of $E_\tau(M)$ containing M. Thus, $Q \cap X = g(Q_\tau(M)) = Q$. Hence, $Q \subseteq X$ and so the restriction $\psi = g|_{Q_\tau(M)}$ is an injective R–linear mapping such that $\psi : Q_\tau(M) \to X$ and $\psi \circ \varphi = f$.

Finally, we show that $Q_\tau(M)$ is unique up to isomorphism. If $\varphi' : M \to Q$ is also a τ–quasi– injective hull of M, then by the above there are R–linear injections $g : Q_\tau(M) \to Q$ and $h : Q \to Q_\tau(M)$ such that the diagram

$$\begin{array}{ccc} & & Q_\tau(M) \\ & \varphi \nearrow & \downarrow g \\ M & \xrightarrow{\varphi'} & Q \\ & \varphi \searrow & \downarrow h \\ & & Q_\tau(M) \end{array}$$

is commutative where φ is the canonical injection. We claim that $h \circ g$ is the identity map on $Q_\tau(M)$. If $(h \circ g - 1_{Q_\tau(M)})(Q_\tau(M)) \neq 0$, then $(h \circ g - 1_{Q_\tau(M)})(Q_\tau(M)) \cap M \neq 0$ since M is essential in $Q_\tau(M)$. Suppose $0 \neq x \in (h \circ g - 1_{Q_\tau(M)})(Q_\tau(M)) \cap M$. Then $(h \circ g - 1_{Q_\tau(M)})(x) =$ $h \circ g(x) - 1_{Q_\tau(M)}(x) = h \circ g \circ \varphi(x) - x = \varphi(x) - x = x - x = 0$. Thus, $x = 0$ since $h \circ g - 1_{Q_\tau(M)}$ is an injective mapping. But this contradicts the fact that $x \neq 0$ and so $h \circ g(x) = 1_{Q_\tau(M)}$. Similarly, $g \circ h = 1_Q$ and so $Q_\tau(M)$ is unique up to isomorphism. □

When τ is the torsion theory in which every module is torsion, $Q_\tau(M)$ is the quasi–injective hull $Q(M)$ of Johnson and Wong. Moreover, $Q_\tau(M)$ is the smallest τ–quasi–injective module in $E_\tau(M)$ which contains M.

In §4, we called an R–module M $\overset{f}{\tau}$–quasi–injective if whenever L and N are submodules of M with $L \in F_\tau(N)$, each map $f \in \mathrm{Hom}_R(L, M)$ can be extended to a map $g \in \mathrm{Hom}_R(N, M)$. It is an open question as to whether or not an arbitrary R–module has a $\overset{f}{\tau}$–quasi–injective hull. Some work on this question appears in [5] under the assumption that R is a ring such that $\Omega(M)$ is a filter[1] for every R–module M. Such rings do exist, for example this is the case when R is a valuation domain. In these rings the ideals are linearly ordered and it follows from this observation that $\Omega(M)$ is a filter for every R–module M.

5.2 (Quasi–)Projective Covers Relative to a Torsion Theory

Definition 5.2.1 A mapping $f : M \to N$ is *minimal* if $\ker f$ is small in M, *free* if $\ker f \in \mathscr{F}$ and τ–*minimal* if $\ker f$ is small and in $\mathscr{F}$. A τ–projective module $P_\tau(M)$ together with a τ–minimal epimorphism $\varphi : P_\tau(M) \to M$ is a τ–*projective cover* of M.

[1] See 4.1.15 for the definition of $\Omega(M)$. Here the term filter is used for a collection of right ideals of R which is closed under finite intersections and under extensions by right ideals of R.

Obviously, when τ is the torsion theory in which every module is torsionfree, τ–projective covers are just the projective covers of Bass.

Proposition 5.2.2 *Should it exist, a τ–projective cover of an R–module is unique up to isomorphism.*

Proof. Let $\varphi : P_\tau(M) \to M$ and $\varphi^*: P_\tau(M)^* \to M$ be τ–projective covers of M. Since φ^* is free, there is a map $\psi : P_\tau(M) \to P_\tau(M)^*$ such that $\varphi^* \circ \psi = \varphi$. Hence, $P_\tau(M)^* = \text{Image}\,\psi + \ker \varphi^*$ and so ψ is an epimorphism since φ^* is minimal. But $\ker \psi \subseteq \ker \varphi$ and so $\ker \psi$ is small in $P_\tau(M)$ since submodules of small submodules are small. Hence, ψ is τ–minimal. Consequently, by Proposition 4.2.4, ψ is an isomorphism. □

We now show that τ–projective covers exist when projective covers exist.

Proposition 5.2.3 *If $\pi : P \to M$ is a projective cover of M, then $\varphi : P/t_\tau(\ker \pi) \to M$ is a τ–projective cover of M where φ is the map induced by π.*

Proof. Since $\ker \varphi = \ker \pi / t_\tau(\ker \pi)$, φ is free. We also see, by our observations immediately proceeding Proposition 4.2.4, that $\ker \pi / t_\tau(\ker \pi)$ is small in $P/t_\tau(\ker \pi)$. Hence, φ is a τ–minimal epimorphism. That $P/t_\tau(\ker \pi)$ is τ–projective now follows directly from Proposition 4.2.5. □

Recall that a ring R is right perfect if every R–module has a projective cover. This brings up the question as to what type of ring is characterized by the universal existence of τ–projective covers? We now show $R/t_\tau(R)$ is a right perfect ring if and only if every R–module has a τ–projective cover. This result is an immediate corollary of the following proposition, due to Rangaswamy [33]. Rangaswamy's result establishes a link between the existence of the τ–projective cover of an R–module M and the existence of the projective cover of $M/Mt_\tau(R)$ as an $R/t_\tau(R)$–module.

Proposition 5.2.4 *An* R*–module* M *has a* τ*–projective cover if and only if* $M/Mt_\tau(R)$ *has a projective cover as an* $R/t_\tau(R)$*–module.*

Proof. If M has a τ–projective cover, there is a short exact sequence $0 \to K \to P_\tau(M) \overset{\varphi}{\to} M \to 0$ with $P_\tau(M)$ τ–projective and φ τ–minimal. But then $0 \to (K + P_\tau(M)t_\tau(R))/P_\tau(M)t_\tau(R) \to P_\tau(M)/P_\tau(M)t_\tau(R) \to M/Mt_\tau(R) \to 0$ is exact where, by Proposition 4.2.10, $P_\tau(M)/P_\tau(M)t_\tau(R)$ is a projective $R/t_\tau(R)$–module. But since K is small in $P_\tau(M)$, $(K + P_\tau(M)t_\tau(R))/P_\tau(M)t_\tau(R)$ is small in $P_\tau(M)/P_\tau(M)t_\tau(R)$. Hence, $P_\tau(M)/P_\tau(M)t_\tau(R) \to M/Mt_\tau(R) \to 0$ is a projective cover of $M/Mt_\tau(R)$ as an $R/t_\tau(R)$–module.

Conversely, let $\pi : P \to M/Mt_\tau(R)$ b a projective cover of $M/Mt_\tau(R)$ as an $R/t_\tau(R)$–module. By Proposition 4.2.8, there is a short exact sequence of R–modules $0 \to K \to C \overset{\varphi}{\to} M \to 0$ where C is τ–projective and $K \in \mathscr{F}$. Consider the row exact diagram

$$
\begin{array}{ccccccccc}
 & & & & & & & & 0 \\
 & & & & & & & \nearrow & \\
 & & & & P & \overset{1_P}{\rightarrow} & P & & \\
 & & & & \downarrow h & \nearrow g & \downarrow \pi & & \\
0 & \rightarrow & (K + Ct_\tau(R))/Ct_\tau(R) & \rightarrow & C/Ct_\tau(R) & \overset{\lambda}{\rightarrow} & M/Mt_\tau(R) & \rightarrow & 0 \\
 & & & \nearrow & & & & & \\
 & & \ker g & & & & & & \\
 & \nearrow & & & & & & & \\
0 & & & & & & & &
\end{array}
$$

where λ is induced by the map φ. Now, by Proposition 4.2.10, $C/Ct_\tau(R)$ is a projective $R/t_\tau(R)$–module, so there is an R–linear mapping $g : C/Ct_\tau(R) \rightarrow P$ such that $\pi \circ g = \lambda$. g is an epimorphism because $\ker \pi$ is small in P. Since P is projective, the short exact sequence $0 \rightarrow \ker g \rightarrow C/Ct_\tau(R) \overset{g}{\rightarrow} P \rightarrow 0$ splits. This produces an injective R–linear mapping $h : P \rightarrow C/Ct_\tau(R)$ such that $g \circ h = 1_P$ and submodules X and Y of C such that $C/Ct_\tau(R) = X/Ct_\tau(R) \oplus Y/Ct_\tau(R)$ with $\ker g = X/Ct_\tau(R)$ and $h(P) = Y/Ct_\tau(R)$. But $X/Ct_\tau(R) \subseteq (K + Ct_\tau(R))/Ct_\tau(R)$ and so $(K + Ct_\tau(R))/Ct_\tau(R) = X/Ct_\tau(R) \oplus L/Ct_\tau(R)$ where $L/Ct_\tau(R) = Y/Ct_\tau(R) \cap (K + Ct_\tau(R))/Ct_\tau(R)$. Note also that $L/Ct_\tau(R)$ small in $C/Ct_\tau(R)$ since $(K + Ct_\tau(R))/Ct_\tau(R)$ is small in $C/Ct_\tau(R)$. Also since $K \in \mathscr{F}$ and $Ct_\tau(R) \in \mathscr{T}$, $K \cap Ct_\tau(R) = 0$ gives $K + Ct_\tau(R) = K \oplus Ct_\tau(R)$ so that $X = X' \oplus Ct_\tau(R)$ and $L = L' \oplus Ct_\tau(R)$, where $X' = X \cap K$ and $L' = L \cap K$. Moreover, $K + Ct_\tau(R) = X + L$ implies that $K = X' + L' = X' \oplus L'$. Let $P_\tau(M) = C/X'$ and $K^* = K/X'$. Then $P_\tau(M)t_\tau(R) = (Ct_\tau(R) + X')/X' = X/X'$ and so $P_\tau(M)/P_\tau(M)t_\tau(R) \cong C/X \cong P$ is a projective $R/t_\tau(R)$–module. Thus, by Proposition 4.2.10, $P_\tau(M)$ is a τ–projective R–module. Note also that $P_\tau(M)/K^* \cong C/K \cong M$. Finally,

$K^* = K/X' = L' \in \mathscr{F}$ and K^* is small in $P_\tau(M)$ because K is small in C. Thus, $0 \to K^* \to P_\tau(M) \to M \to 0$ is a τ–projective cover of M. □

Corollary 5.2.5 *Every R–module has a τ–projective cover if and only if $R/t_\tau(R)$ is a right perfect ring.*

Recall that for a torsion theory τ on Mod–R, $\tau_\otimes$ is the torsion theory $(\mathscr{D}, \mathscr{R})$ on R–Mod where $\mathscr{D} = \{ M \in R\text{–Mod} \mid N \otimes_R M = 0 \text{ for all } N \in \mathscr{T} \}$. Note also that since $t_\tau(R) \otimes_R t_{\tau_\otimes}(R) = 0$, $t_\tau(R)t_{\tau_\otimes}(R) = 0$ and so $t_\tau(R)$ is an $R/t_{\tau_\otimes}(R)$–module.

In Proposition 4.3.16 it was shown that when $\tau_\otimes$ is hereditary if the class of τ–flat left R–modules coincides with the class of $\tau_\otimes$–projective left R–modules, then $R/t_{\tau_\otimes}(R)$ is a left perfect ring. The following proposition will complete the tools necessary to prove that under appropriate conditions the converse of this result holds.

Proposition 5.2.6 *Let τ be a torsion theory on* Mod–R *such that $\tau_\otimes$ is hereditary. If M is a τ–flat left R–module, then given any exact sequence $0 \to K \to P \to M/t_{\tau_\otimes}(R)M \to 0$ where $K \in \mathscr{R}$ and P is a $\tau_\otimes$–projective left R–module, there is for each $x \in t_\tau(R)K$ an R–linear mapping $\theta_x : P \to K$ which leaves x fixed.*

Proof. Let $0 \to K \to P \to M/t_{\tau_\otimes}(R)M \to 0$ be exact where $K \in \mathscr{R}$, P is a $\tau_\otimes$–projective left R–module and M is a τ–flat left R–module. Since $t_{\tau_\otimes}(R)K \subseteq t_{\tau_\otimes}(R)P \subseteq K \in \mathscr{R}$ and $t_{\tau_\otimes}(R)K, t_{\tau_\otimes}(R)P \in \mathscr{D}$, $t_{\tau_\otimes}(R)K = t_{\tau_\otimes}(R)P = 0$. Hence, the sequence of left R–modules $0 \to K \to P \to M/t_{\tau_\otimes}(R)M \to 0$ can be viewed as an exact sequence of left $R/t_{\tau_\otimes}(R)$–modules. Moreover, Proposition 4.2.10 shows that P is a projective left $R/t_{\tau_\otimes}(R)$–module. Thus, there exists a projective left $R/t_{\tau_\otimes}(R)$–module P' such that $P \oplus P'$ is a free left $R/t_{\tau_\otimes}(R)$–module and such that the sequence $0 \to K \to P \oplus P' \to M/t_{\tau_\otimes}(R)M \oplus P' \to 0$ is exact. Since P' is a projective left $R/t_{\tau_\otimes}(R)$–module, Proposition 4.2.10 also shows that P' is a $\tau_\otimes$–projective left R–module and consequently a τ–flat left R–module. Since M is a τ–flat left R–module, Corollary 4.3.17 indicates that $M/t_{\tau_\otimes}(R)M$ is a τ–flat left R–module. Hence, $M/t_{\tau_\otimes}(R)M \oplus P'$ is a τ–flat left R–module.

Let $\{ b_\alpha \}_{\alpha \in \Delta}$ be a basis for free left $R/t_{\tau_\otimes}(R)$–module $P \oplus P'$, let $t \in t_\tau(R)$ and suppose that $k \in K$. If $k = \sum_{i=1}^{n} r_i b_{\alpha_i}$ in $P \oplus P'$, set $I = \sum_{i=1}^{n} r_i R$, so that $t_\tau(R)/tI \in \mathcal{T}$. Since $M/t_{\tau_\otimes}(R)M \oplus P'$ is a τ–flat left R–module, it follows from the fact that $\mathrm{Tor}_1^R(\, t_\tau(R)/tI\, , M/t_{\tau_\otimes}(R)M \oplus P'\,) = 0$ that the sequence $0 \to t_\tau(R)/tI \otimes_R K \to t_\tau(R)/tI \otimes_R (\, P \oplus P'\,) \to t_\tau(R)/tI \otimes_R (\, M/t_{\tau_\otimes}(R)M \oplus P'\,) \to 0$ is exact. In $t_\tau(R)/tI \otimes_R (\, P \oplus P'\,)$, we have $(\, t + tI\,) \otimes k = (\, t + tI\,) \otimes \sum_{i=1}^{n} r_i b_{\alpha_i}$ $= \sum_{i=1}^{n} (tr_i + tI\,) \otimes b_{\alpha_i} = \sum_{i=1}^{n} (\, 0 \otimes b_{\alpha_i}\,) = 0$. Thus $(\, t + tI\,) \otimes k = 0$ in $t_\tau(R)/tI \otimes_R K$. Under the canonical map $t_\tau(R)/tI \otimes_R K \to R/tI \otimes_R K$, we have $(\, t + tI\,) \otimes k \to (\, t + tI\,) \otimes k = (\, 1 + tI\,) \otimes tk$. Hence, $(\, 1 + tI\,) \otimes tk = 0$ in $R/tI \otimes_R K$. Next, consider the sequence $tI \otimes_R K \to R \otimes_R K \to R/tI \otimes_R K \to 0$. Since $1 \otimes tk$ in $R \otimes_R K$ maps to $0 = (\, 1 + tI\,) \otimes tk$ in $R/tI \otimes_R K$, $1 \otimes tk$ is the image of an element in $tI \otimes_R K$, say $1 \otimes tk = \sum_{i=1}^{n} (\, tr_i \otimes k_i\,)$. Hence, we see that $tk = \sum_{i=1}^{n} t\, r_i\, k_i = t \sum_{i=1}^{n} r_i\, k_i$. Now define $\phi : P \oplus P' \to K$ by $\phi(\, b_{\alpha_i}) = k_i$ for $i = 1, 2, \ldots, n$ and $\phi(\, b_\alpha\,) = 0$ if $\alpha \notin \{\, \alpha_1, \alpha_2, \ldots, \alpha_n\,\}$. Then $\phi(\, tk\,) = \phi(\, t \sum_{i=1}^{n} r_i b_{\alpha_i}) = t \sum_{i=1}^{n} r_i \phi(\, b_{\alpha_i}) = t \sum_{i=1}^{n} r_i k_i = tk$ and so if we define $\theta = \phi|_P$, $\theta : P \to K$ is such that $\theta(\, tk\,) = tk$.

Next, let $t_1k_1, t_2k_2, \ldots, t_nk_n \in t_\tau(R)K$ where $t_i \in t_\tau(R)$ and $k_i \in K$ for $i = 1, 2, \ldots, n$. We claim for any such collection of elements of $t_\tau(R)K$ there is an R–linear mapping $\theta : P \to K$ which will leave t_ik_i fixed for $i = 1, 2, \dots, n$. We proceed by induction. If $n = 1$, the result follows from the proceeding paragraphs of this proof. Next, assume $n = m$ and make the induction hypothesis that for any elements $t_1k_1, t_2k_2, \dots, t_mk_m$ in $t_\tau(R)K$ there is an R–linear mapping $\theta : P \to K$ which leaves t_ik_i fixed for $i = 1, 2, \dots, m$. Now suppose $t_1k_1, t_2k_2, \dots, t_{m+1}k_{m+1} \in t_\tau(R)K$. We know from the case for $n = 1$, that there is an R–linear mapping $\theta' : P \to K$ which leaves $t_{m+1}k_{m+1}$ fixed. Since $t_i\,(k_i - \theta'(k_i)) \in t_\tau(R)K$ for $i = 1, 2, \dots, m$, by the induction hypothesis, there is a map $\theta^*: P \to K$ which leaves theses elements fixed. Now consider $\theta = 1_P - (1_P - \theta^*)(1_P - \theta')$. Then $\theta(t_ik_i) = t_ik_i - (1 - \theta^*)(1 - \theta')(t_ik_i) = t_ik_i - (1 - \theta^*)(t_i\,(k_i - \theta'(k_i)) = t_ik_i$ for $i = 1, 2, \dots, m$ and $\theta(t_{m+1}k_{m+1}) = t_{m+1}k_{m+1} - (1 - \theta^*)(1 - \theta')(t_{m+1}k_{m+1})$ $= t_{m+1}k_{m+1}$. Therefore, by induction, if $x \in t_\tau(R)K$ and $x = \sum_{i=1}^{n} t_i\,k_i$, there is an R–linear mapping $\theta_x : P \to K$ which leaves t_ik_i fixed for each i. Consequently, $\theta_x(x) = x$. □

Now for the converse of Proposition 4.3.16.

Proposition 5.2.7 *Let τ be a torsion theory such that $\tau_\otimes$ is hereditary. If $t_\tau(R)$ is a faithful left $R/t_{\tau_\otimes}(R)$–module and $R/t_{\tau_\otimes}(R)$ is a left perfect ring, then the class of $\tau_\otimes$–projective left R–modules coincides with the class of τ–flat left R–modules.*

Proof. Assume that $R/t_{\tau_\otimes}(R)$ is a left perfect ring and let M be a τ–flat left R–module. Then Corollary 5.2.5 shows that when $M/t_{\tau\otimes}(R)M$ is viewed as left R–module it has a $\tau_\otimes$–projective cover $\varphi : P \to M/t_{\tau\otimes}(R)M$ where $K = \ker\varphi \in \mathscr{R}$ and K is small in P. By Proposition 5.2.6, if $x \in t_\tau(R)K$, there is an R–linear mapping $\theta_x : P \to K$ such that $\theta_x(x) = x$. We claim that $1_P - \theta_x : P \to P$ is an isomorphism. First, note that if $(1_P - \theta_x)(y) = 0$, then $y = \theta_x(y) \in K$ and so $\ker(1_P - \theta_x) \in \mathscr{R}$ and $\ker(1_P - \theta_x)$ is small in P. We also see that $\operatorname{Im}(1_P - \theta_x) = P$ since $\operatorname{Im}(\theta_x) + \operatorname{Im}(1_P - \theta_x) = P$ and $\operatorname{Im}(\theta_x) \subseteq K$ is small in P. Hence, $1_P - \theta_x$ is an epimorphism. Now consider the row exact commutative diagram

$$\begin{array}{ccccccccc} & & & & & & P & & \\ & & & & & \psi \swarrow & \downarrow 1_P & & \\ 0 & \to & \ker(1_P - \theta_x) & \to & P & \xrightarrow{1_P - \theta_x} & P & \to & 0 \end{array}$$

where ψ is given by the $\tau_\otimes$–projectivity of P. Since ψ is an epimorphism and $(1_P - \theta_x) \circ \psi = 1_P$, $1_P - \theta_x$ is a monomorphism and so is an isomorphism. Next, note that $t_\tau(R)K = 0$, for if $x \in t_\tau(R)K$, then $(1_P - \theta_x)(x) = x - x = 0$. Thus, $x = 0$ since $1_P - \theta_x$ is an isomorphism.

Since $t_{\tau_\otimes}(R)P \subseteq K \in \mathscr{R}$ and $t_{\tau_\otimes}(R)P \in \mathscr{D}$, $t_{\tau_\otimes}(R)P = 0$ and so P is a left $R/t_{\tau_\otimes}(R)$–module. Thus, by Proposition 4.2.10, P is a projective left $R/t_{\tau_\otimes}(R)$–module and so for some indexing set Δ, there is an embedding of P into $\oplus_{\alpha \in \Delta}(R/t_{\tau_\otimes}(R))_\alpha$. If P is identified with its image in $\oplus_{\alpha \in \Delta}(R/t_{\tau_\otimes}(R))_\alpha$ and $k \in K$, then k can be written as $(r_\alpha + t_{\tau_\otimes}(R))_{\alpha \in \Delta}$. But $t_\tau(R)K = 0$ implies $t_\tau(R)(r_\alpha + t_{\tau_\otimes}(R)) = 0$ for each $\alpha \in \Delta$ and so since it was assumed that $t_\tau(R)$ is a faithful $R/t_{\tau_\otimes}(R)$–module, $r_\alpha + t_{\tau_\otimes}(R) = 0$ for each $\alpha \in \Delta$. Hence, $k = 0$. Consequently, $K = 0$ and so $M/t_{\tau_\otimes}(R)M \cong P$. But since $M/t_{\tau_\otimes}(R)M$ is a projective left $R/t_{\tau_\otimes}(R)$–module, it now follows from Proposition 4.2.10 that M is a $\tau_\otimes$–projective left R–module. Thus, every τ–flat left R–module is $\tau_\otimes$–projective. Since Proposition 4.4.13 shows that any $\tau_\otimes$–projective left R–module is τ–flat, we have the proposition. □

Rings such that $R/t_\tau(R)$ is right perfect rings can also be characterized using τ–quasi–projective covers [7].

Definition 5.2.8 A τ–quasi–projective module $QP_\tau(M)$ together with a τ–minimal epimorphism $\varphi : QP_\tau(M) \to M$ is said to be a *τ–quasi–projective cover* of M if whenever $0 \neq K \subseteq \ker \varphi$, $QP_\tau(M)/K$ is not τ-quasi-projective.

Proposition 5.2.9 *Suppose that* τ *is a torsion theory on* Mod–R *and let* $\varphi : P \to M$ *be an epimorphism where* P *is* τ*–projective and* $\ker \varphi \in \mathcal{F}$*. If* $\ker \varphi$ *is stable under endomorphisms of* P*, then* M *is* τ*–quasi–projective.*

Proof. Consider the row exact diagram

$$\begin{array}{ccccccc} & & & & P & & \\ & & & & \downarrow \varphi & & \\ & & h \swarrow & & M & & \\ & & & h^* \swarrow & \downarrow & & \\ P & \overset{\varphi}{\to} & M & \overset{f}{\to} & N & \to & 0 \end{array}$$

where $\ker f \in \mathcal{F}$. If $K = \ker f \circ \varphi$, then $f \circ \varphi(\, t_\tau(K)\,) = 0$ and so $\varphi(\, t_\tau(K) \subseteq \ker f$. Hence, it follows that $\varphi(\, t_\tau(K)\,) = 0$ and so $t_\tau(K) \subseteq \ker \varphi$. Consequently, $t_\tau(K) = 0$ and so K is in $\mathcal{F}$. Thus, the τ–projectivity of P ensures the existence of an R–linear mapping $h : P \to P$ which makes the outer diagram commute. But $h(\, \ker \varphi\,) \subseteq \ker \varphi$ and so we have an induced map $h^* : M \to M$ which renders the inner diagram commutative. Hence, M is τ–quasi–projective. $\square$

Proposition 5.2.10 *If* τ *is cohereditary and* $\varphi : P_\tau(M) \to P_\tau(M)/K : x \to x + K$ *is a* τ*–projective cover of the* τ*-quasi-projective module* $P_\tau(M)/K$*, then* K *is stable under endomorphisms of* $P_\tau(M)$.

Proof. Let f be an endomorphism of $P_\tau(M)$. Since the mapping $K \to f(K) : x \to f(x)$ is an epimorphism and $K \in \mathscr{F}$, it follows that $f(K) \in \mathscr{F}$. Thus, since $K \oplus f(K) \to K + f(K) : (x, y) \to x + y$ is an epimorphism, $K + f(K) \in \mathscr{F}$. Now f induces a map $f^*: P_\tau(M)/K \to P_\tau(M)/(K + f(K))$ given by $f^*(x + K) = f(x) + K + f(K)$ and so consider the diagram

$$\begin{array}{ccc} & & P_\tau(M)/K \\ & \beta \swarrow & \downarrow f^* \\ P_\tau(M)/K & \stackrel{\eta}{\to} & P_\tau(M)/(K + f(K)) \to 0 \end{array}$$

where $\eta(x + K) = x + K + f(K)$. The diagram can be completed commutatively by a map β, since $P_\tau(M)/K$ is τ-quasi-projective and $\ker \eta \in \mathscr{F}$. Thus, by using the τ–projectivity of $P_\tau(M)$, we have a commutative diagram

$$\begin{array}{ccc} P_\tau(M) & \stackrel{\varphi}{\to} & P_\tau(M)/K \\ \downarrow \alpha & & \downarrow \beta \\ P_\tau(M) & \stackrel{\varphi}{\to} & P_\tau(M)/K \to 0 \end{array}$$

Next, let $X = \{ x \in P_\tau(M) \mid f(x) - \alpha(x) \in K \}$. We claim that $X = P_\tau(M)$. Since $\varphi \circ \alpha(K) = \beta \circ \varphi(K) = 0$, $\alpha(K) \subseteq \ker \varphi = K$ and so α induces a map $\alpha^*: P_\tau(M)/K \to P_\tau(M)/(K + f(K)) : x + K \to \alpha(x) + K + f(K)$. Hence,

$$\begin{aligned}(f^* - \alpha^*)(x+K) &= f^*(x+K) - \alpha^*(x+K)\\ &= \eta \circ \beta(x+K) - (\alpha(x) + K + f(K))\\ &= \eta \circ \beta(x+K) - \eta \circ \varphi \circ \alpha(x)\\ &= \eta \circ \beta(x+K) - \eta \circ \beta \circ \varphi(x)\\ &= \eta \circ \beta(x+K) - \eta \circ \beta(x+K)\\ &= 0.\end{aligned}$$

Thus, $f^*(x+K) - \alpha^*(x+K) = 0$ and, therefore, $f(x) + K + f(K) - (\alpha(x) + K + f(K)) = 0$. Consequently, $f(x) - \alpha(x) \in K + f(K)$. Now let $f(x) - \alpha(x) = k_1 + f(k_2)$, $k_1, k_2 \in K$. Then $f(x - k_2) - \alpha(x - k_2) = k_1 + \alpha(k_2) \in K$, since $\alpha(k_2) \in \alpha(K) \subseteq K$. Thus, $x - k_2 \in X$ and so $P_\tau(M) = K + X$. But K is small in $P_\tau(M)$ and so $P_\tau(M) = X$. Hence, if $x \in K$, then $x \in X$ and so $f(x) - \alpha(x) \in K$. But $\alpha(x) \in K$ and, therefore, $f(x) \in K$. This shows that $f(K) \subseteq K$. □

Proposition 5.2.11 *If* τ *is coheredítary and if* M *has a* τ*–projective cover* $\varphi : P_\tau(M) \to M$, *then* M *has a* τ*–quasi–projective cover* $\psi : QP_\tau(M) \to M$ *which is unique up to isomorphism.*

Proof. Via Zorn's lemma there is a unique submodule X of $K = \ker \varphi$ maximal with respect to the property that X is stable under endomorphisms of $P_\tau(M)$. Set $QP_\tau(M) = P_\tau(M)/X$. If $\psi : QP_\tau(M) \to M$ is the epimorphism induced by φ, then $\ker \psi = K/X$. Now K is small in $P_\tau(M)$ and so $\ker \psi$ is small in $QP_\tau(M)$. Note also that $\ker \psi \in \mathscr{F}$ since τ is cohereditary. Hence, we have a map $P_\tau(M) \to P_\tau(M)/X$ where $P_\tau(M)$ is τ–projective, $X \in \mathscr{F}$ and X is stable under endomorphisms of $P_\tau(M)$. Then, by Proposition 5.2.9, $QP_\tau(M)$ is τ-quasi-projective. Next, let $Y/X \subseteq \ker \psi$ be such that

$(P_\tau(M)/X)/(Y/X) \cong P_\tau(M)/Y$ is τ-quasi-projective where $X \subseteq Y \subseteq K$. Then $0 \to Y \to P_\tau(M) \to P_\tau(M)/Y \to 0$ is a τ–projective cover of $P_\tau(M)/Y$. Hence, it follows from Proposition 5.2.10 that Y is stable under endomorphisms of $P_\tau(M)$ and so it must be the case that $X = Y$. Thus, $\psi : QP_\tau(M) \to M$ is a τ–quasi-projective cover of M.

Finally, let's show uniqueness. Suppose $\phi : QP \to M$ is another a τ–quasi–projective cover of M and consider the diagram

$$\begin{array}{ccc} & P_\tau(M) & \\ \mu \swarrow & \downarrow \varphi & \\ QP & \overset{\phi}{\to} & M \to 0 . \end{array}$$

Since $\ker \phi \in \mathscr{F}$ and $P_\tau(M)$ is τ–projective, this diagram can be completed commutatively by a map μ which must be an epimorphism because $\ker \phi$ is small in QP. Now $\ker \mu \subseteq K$ and so it follows that $\mu : P_\tau(M) \to QP$ is a τ–projective cover of QP. Hence, by Proposition 5.2.10, $\ker \mu$ is stable under endomorphisms of $P_\tau(M)$. Thus, if X is as above, then $\ker \mu \subseteq X$. Suppose $\ker \mu \neq X$, then $\mu(X) \neq 0$. Now $\phi \circ \mu(X) = \varphi(X) \subseteq \varphi(K) = 0$ and so $0 \neq \mu(X) \subseteq \ker \phi$. The map $\mu^*: P_\tau(M) \to QP/\mu(X) : y \to \mu(y) + \mu(X)$ is an epimorphism and we claim that $\ker \mu^* = X$. If $y \in X$, then $\mu(y) + \mu(X) = 0$ and so $\mu^*(y) = 0$. Hence, $X \subseteq \ker \mu^*$. If $y \in \ker \mu^*$, then $\mu(y) + \mu(X) = 0$. Let $\mu(y) = \mu(x)$, so that $\mu(y - x) = 0$. Then $y - x \in \ker \mu \subseteq X$ and so we see that $y \in X$ and therefore that $\ker \mu^* \subseteq X$. Hence, $\ker \mu^* = X$ which implies that $QP_\tau(M) = P_\tau(M)/X \cong QP/\mu(X)$. But this contradicts the definition of a τ-quasi-projective cover. Consequently, $\ker \mu = X$ and so $\mu(X) = 0$. This yields $QP_\tau(M) = P_\tau(M)/X \cong QP$ and so $QP_\tau(M)$ is unique up to isomorphism. □

We conclude this section with the following characterization of rings for which $R/t_\tau(R)$ is right perfect rings.

Proposition 5.2.12 *If* τ *is cohereditary, then every* R*–module has a* τ*–quasi–projective cover if and only if* $R/t_\tau(R)$ *is a right perfect ring.*

Proof. Assume that every R–module has a τ-quasi-projective cover. Let M be an $R/t_\tau(R)$–module and suppose that $\theta : F \to M$ is a free $R/t_\tau(R)$–module on M. Next, suppose that $\varphi : QP_\tau(M) \to F \oplus M$ is a τ–quasi-projective cover of $F \oplus M$. Since $(F \oplus M)t_\tau(R) = 0$, $QP_\tau(M)t_\tau(R) \subseteq \ker \varphi$. But $QP_\tau(M)t_\tau(R) \subseteq t_\tau(QP_\tau(M))$ and so $QP_\tau(M)t_\tau(R) = 0$ since $\ker \varphi \in \mathscr{F}$. Thus, $QP_\tau(M)$ is an $R/t_\tau(R)$–module.

Now consider

$$\begin{array}{ccccccc} & & & & F & & \\ & & \swarrow f & & \downarrow 1_F & & \\ QP_\tau(M) & \xrightarrow{\varphi} & F \oplus M & \xrightarrow{p_1} & F & \to & 0 \end{array}$$

where p_1 is the first projection map and f is the completing map given by the projectivity of F. If $M^* = \ker(p_1 \circ \varphi)$, we can assume that $QP_\tau(M) = F \oplus M^*$. Let $\varphi^* = \varphi|_{M^*}$. We claim that $\varphi^* : M^* \to M$ is an $R/t_\tau(R)$–projective cover of M. Clearly φ^* is an epimorphism with small kernel and so consider the diagram

$$\begin{array}{ccccc} & & F & & \\ & \theta^* \swarrow & \downarrow \theta & & \\ M^* & \xrightarrow{\varphi^*} & M & \to & 0 \end{array}$$

which completes commutatively by the projectivity of F. Note that θ^* is an epimorphism since $\ker \varphi^*$ is small in M^*. Hence, we have a diagram

$$\begin{array}{ccccccc} & & & & F \oplus M^* & & \\ & & & & \downarrow p_2 & & \\ & h \swarrow & & & M^* & & \\ & & & h^* \swarrow & \downarrow 1_{M^*} & & \\ F \oplus M^* & \xrightarrow{p_1} & F & \xrightarrow{\theta^*} & M^* & \rightarrow & 0\,. \end{array}$$

Since $0 \rightarrow \ker \varphi^* \rightarrow M^* \xrightarrow{\varphi^*} M \rightarrow 0$ is exact and $\ker \varphi^*$ and M are both in $\mathcal{F}$, it follows that $M^* \in \mathcal{F}$ because $\mathcal{F}$ is closed under extensions. Consequently, $\ker \theta^* \circ p_1 \in \mathcal{F}$ and so we have a completing map h. If $j_2: M^* \rightarrow F \oplus M^*$ is the canonical injection and $h^* = p_1 \circ h \circ j_2$, then the inner diagram is commutative and so M^* is a projective $R/t_\tau(R)$–module. Thus, $\varphi^*: M^* \rightarrow M$ is a $R/t_\tau(R)$–projective cover of M. Therefore, $R/t_\tau(R)$ is a right perfect ring.

Conversely, if $R/t_\tau(R)$ is a right perfect ring, then every $R/t_\tau(R)$–module has a $R/t_\tau(R)$–projective cover. Thus, if M is any R–module, then $M/Mt_\tau(R)$ has an $R/t_\tau(R)$–projective cover. Therefore, by Proposition 5.2.4 , M has a τ–projective cover as an R–module. But then, by Proposition 5.2.11, M has a τ-quasi-projective cover. □

5.3 Torsionfree Covers

Definition 5.3.1 Let τ be a torsion theory on Mod–R. An R–linear mapping $\varphi : M^* \rightarrow M$ is said to satisfy the *torsionfree factor property* if for $X \in \mathcal{F}$ and any R–homomorphism $\phi : X \rightarrow M$ there is an R–linear mapping $f : X \rightarrow M^*$ such that the diagram

$$\begin{array}{ccc} & & X \\ & f \swarrow & \downarrow \phi \\ M^* & \overset{\varphi}{\to} & M \to 0 \end{array}$$

is commutative. $\varphi : M^* \to M$ *is said to have* TFP when $\varphi : M^* \to M$ satisfies the torsionfree factor property. If this condition holds only when X is a τ–torsionfree injective module, then $\varphi : M^* \to M$ *is said to have* TFP *for torsionfree injective modules.* An R–module $T_\tau(M)$ together with an R–linear epimorphism $\varphi : T_\tau(M) \to M$ is a *τ–torsionfree precover* of M if $T_\tau(M) \in \mathscr{F}$ and $\varphi : T_\tau(M) \to M$ has TFP. If a τ–torsionfree precover $\varphi : T_\tau(M) \to M$ also satisfies the property that $\ker \varphi$ contains no nonzero τ–pure submodules of $T_\tau(M)$, then $\varphi : T_\tau(M) \to M$ is said to be a *τ–torsionfree cover* of M. If every R–module has a τ–torsionfree cover, then τ is said to be *universally covering.*

τ–torsionfree covers were first defined by Enochs [12] when τ is the usual torsion theory on Mod–R over an integral domain. Enochs showed that this torsion theory is universally covering. His results were extended to perfect torsion theories by Banaschewski [2]. Additional results on the existence and properties of τ–torsionfree covers were discovered by Enochs [13] and Golan and Teply [21] while Cheatham [9] characterized the left non–singular rings for which the Goldie torsion theory is universally covering. The main result presented here is due to Teply [35]. He showed that a faithful torsion theory τ for which $F_\tau(R)$ contains a cofinal subset of finitely generated right ideals is universally covering. The the first version of the following proposition appears in [12] where τ is the usual torsion theory on Mod–R associated with an integral domain. The more general version presented here can be found in [36].

Proposition 5.3.2 *A τ–torsionfree cover of* M *is unique up to isomorphism whenever it can be shown to exist.*

Proof. Let $\varphi : T_\tau(M) \to M$ and $\varphi' : T_\tau(M)' \to M$ be τ–torsionfree covers of M. By definition there is an R–linear mapping $f : T_\tau(M) \to T_\tau(M)'$ such that $\varphi' \circ f = \varphi$. We claim that $K = \ker f$ is τ–pure in $T_\tau(M)$. If not, $T_\tau(M)/K$ contains a nonzero τ–torsion element $x + K$. But then, there is an element $J \in F_\tau(R)$ such that $xJ \subseteq K$. Hence, $f(x)J = 0$ and so $f(x)$ is a τ–torsion element of $T_\tau(M)'$. Thus, $f(x) = 0$ and so $x \in K$, a contradiction. Therefore, K is τ–pure in $T_\tau(M)$. But this tells us that $K = 0$ since $K \subseteq \ker \varphi$ and $\varphi : T_\tau(M) \to M$ is a τ–torsionfree cover of M. Therefore, f is an injection and so $\mathrm{Card}(T_\tau(M)) \leq \mathrm{Card}(T_\tau(M)')$. Similarly, $\mathrm{Card}(T_\tau(M)') \leq \mathrm{Card}(T_\tau(M))$ and so any two τ–torsionfree covers of M have the same cardinality. Now let X be a set containing the elements of $T_\tau(M)$ and $T_\tau(M)'$ and such that $\mathrm{Card}(X) > \mathrm{Card}(T_\tau(M))$. Let $\mathcal{S}$ be the set of all pairs (T_0, φ_0) where T_0 is an R–module whose elements are elements of X and where $\varphi_0 : T_0 \to M$ is τ–torsionfree cover of M. $\mathcal{S}$ is clearly nonempty since $\varphi : T_\tau(M) \to M$ gives a pair in $\mathcal{S}$. Partial order $\mathcal{S}$ by $(T_0, \varphi_0) \leq (T_1, \varphi_1)$ if and only if $T_0 \subseteq T_1$ and $\varphi_1 |_{T_0} = \varphi_0$. If $\mathcal{C}$ is a chain in $\mathcal{S}$, let T^* be the R–module which results from taking the union of the first components of the elements of $\mathcal{C}$ and let $\varphi^* : T^* \to M$ be the R–linear mapping defined by $\varphi^* |_{T_0} = \varphi_0$ for each pair (T_0, φ_0) in $\mathcal{C}$. Then $\varphi^* : T^* \to M$ clearly has TFP. If K is a τ–pure submodule of T^* contained in $\ker \varphi^*$, then $K \cap T_0$ is a τ–pure submodule of T_0 contained in $\ker \varphi_0$ for each pair $(T_0, \varphi_0) \in \mathcal{C}$. From this it follows that $K = 0$ and so $(T^*, \varphi^*) \in \mathcal{S}$. Hence, (T^*, φ^*) is an upper bound for $\mathcal{C}$. Via Zorn's lemma choose (T^*, φ^*) to be a maximal element of $\mathcal{S}$ and let $f_1 : T^* \to T_\tau(M)$ be an R–linear mapping such that $\varphi \circ f_1 = \varphi^*$. By an argument similar to that in the first part of this proof, we can show that f_1 is a monomorphism. We claim that f_1 is also an epimorphism. To see this, let $Y \subseteq X$ be such that $\mathrm{Card}(Y) = \mathrm{Card}(\, T_\tau(M)\,) \setminus f_1(T^*)\,)$ and such that $T^* \cap Y = \varnothing$. Such a Y exists since $\mathrm{Card}(X) > \mathrm{Card}(\, T_\tau(M)\,) = \mathrm{Card}(T^*)$. Next, let $T_0 = T^* \cup Y$ and let $g : T_0 \to T_\tau(M)$ be a bijection such that $g\,|_{T^*} = f_1$ and $g(Y) = T_\tau(M) \setminus f_1(T^*)$. Make T_0 into an R–module as follows: If $x, y \in T_0$ and $r \in R$, set $x + y = g^{-1}(\, g(x) + g(y)\,)$ and $xr = g^{-1}(\, g(x)r\,)$. With this R–module structure $g : T_0 \to T_\tau(M)$ is an R–linear isomorphism. Moreover, $T^* \subseteq T_0$, $(\, T_0, \varphi \circ g\,)$ is such that $\varphi \circ g\,|_{T^*} = \varphi \circ f_1 = \varphi^*$ and $(\, T_0, \varphi \circ g\,) \in \mathcal{S}$. Hence, $(\, T^*, \varphi^*\,) \leq (\, T_0, \varphi \circ g\,)$. But $(\, T^*, \varphi^*\,)$ is maximal and so $T^* = T_0$. Thus, $Y = \varnothing$ and so $f_1(T^*) = T_\tau(M)$. Therefore, f_1 is an epimorphism. Similarly, any linear mapping $f_2 : T^* \to T_\tau(M)'$ such that $\varphi' \circ f_2 = \varphi^*$ must be an epimorphism. But $f \circ f_1$ is such a mapping since $\varphi' \circ f \circ f_1 = \varphi \circ f_1 = \varphi^*$. Hence, $f \circ f_1$ is an epimorphism and so f is an epimorphism and, consequently, an isomorphism. □

Recall that $F_\tau(R)$ contains a cofinal subset of finitely generated right ideals of R if for every $K \in F_\tau(R)$, there is a finitely generated right ideal J of R such that $J \in F_\tau(R)$ and $J \subseteq K$.

Proposition 5.3.3 *If* $F_\tau(R)$ *has a cofinal subset of finitely generated right ideals, then the union of a chain of* τ*–pure submodules of an* R*–module* M *is* τ*–pure in* M.

Proof. Suppose that $\{\, N_\alpha\}_{\alpha \in \Delta}$ is a chain of τ–pure submodules of M. If $\cup_{\alpha \in \Delta} N_\alpha$ is not τ–pure in M, then $M/\cup_{\alpha \in \Delta} N_\alpha \notin \mathcal{F}$. Hence, there must exist an nonzero τ–torsion element $x + \cup_{\alpha \in \Delta} N_\alpha \in M/\cup_{\alpha \in \Delta} N_\alpha$ and a $K \in F_\tau(R)$ such that $\left(x + \cup_{\alpha \in \Delta} N_\alpha\right)K = 0$. Thus, $xK \subseteq \cup_{\alpha \in \Delta} N_\alpha$. But there is a finitely generated right ideal J of R such that $J \in F_\tau(R)$ and $J \subseteq K$ and so we have $xJ \subseteq \cup_{\alpha \in \Delta} N_\alpha$. It follows from the fact that J is finitely generated that there is a $\beta \in \Delta$ such that $xJ \subseteq N_\beta$. Since β can also be chosen so that $x + N_\beta \neq 0$ in M/N_β, $x + N_\beta$ is a nonzero τ–torsion element of M/N_β. Hence, N_β is not τ–pure in M which is a contradiction. Therefore, $\cup_{\alpha \in \Delta} N_\alpha$ must be τ–pure in M. □

The following proposition represents an adaption to a general torsion theoretical setting of a sequence of lemmas originally due to Enochs [12].

Proposition 5.3.4 *Let* τ *be a torsion theory on* Mod R. *If* M *is an* R*–module, then:*

1. *If* $\varphi : M^* \to M$ *has* TFP *and* $N \subseteq M$, *then* $\varphi|_{\varphi^{-1}(N)} : \varphi^{-1}(N) \to N$ *has* TFP.
2. *If* M *is injective, then* $\varphi : M^* \to M$ *has* TFP *if and only if* $\varphi : M^* \to M$ *has* TFP *for torsionfree injective modules.*
3. *If* $\varphi : M^* \to M$ *has* TFP *and* N *is a submodule of* M^* *contained in* $\ker \varphi$, *then the induced map* $\bar{\varphi} : M^*/N \to M$ *has* TFP.

Proof. 1. Suppose $X \in \mathscr{F}$ and let $\phi : X \to N$ be an R–linear mapping. Since $\phi : X \to N \subseteq M$ and $\varphi : M^* \to M$ has TFP, there is an R–linear mapping $f : X \to M$ such that $\varphi \circ f = \phi$. But $\varphi \circ f(X) = \phi(X) \subseteq N$ and so $f(X) \subseteq \varphi^{-1}(N)$. Thus, $\varphi|_{\varphi^{-1}(N)} \circ f = \phi$ which indicates that $\varphi|_{\varphi^{-1}(N)} : \varphi^{-1}(N) \to N$ has TFP.

2. Suppose M is an injective R–module. If $\varphi : M^* \to M$ has TFP, then the condition obviously follows. Conversely, suppose $\phi : X \to M$ is an R–linear mapping where X is τ–torsionfree. By assumption τ is hereditary and so the injective hull E(X) of X is in $\mathscr{F}$. Since M is injective, there is

an R–linear mapping $\phi_E : E(X) \to M$ such that $\phi_E|_X = \phi$ where we assume, without loss of generality, that $X \subseteq E(X)$. Therefore, if the condition holds, there is an R–linear mapping $f : E(X) \to M^*$ such that $\varphi \circ f = \phi_E$. But then $\varphi \circ (f|_X) = \phi_E|_X = \phi$ and so $\varphi : M^* \to M$ has TFP.

3. Suppose $\varphi : M^* \to M$ has TFP and let N be a submodule of M^* contained in ker φ. If $\phi : X \to M$ is an R–linear mapping where $X \in \mathscr{F}$, then there is an R–linear mapping $f : X \to M^*$ such that $\varphi \circ f = \phi$. Since $\bar{\varphi} \circ \pi = \varphi$ where $\pi : M^* \to M^*/N$ is the canonical surjection, we see that $\pi \circ f : X \to M^*/N$ is such that $\bar{\varphi} \circ (\pi \circ f) = \phi$. Hence, $\bar{\varphi} : M^*/N \to M$ has TFP. □

Proposition 5.3.5 *Let* τ *be a faithful torsion theory on* Mod–R *and suppose that* $F_\tau(R)$ *contains a cofinal subset of finitely generated right ideals of* R. *If* M *is an injective* R*–module, then* $\varphi : T_\tau(M) \to M$ *is a* τ*–torsionfree precover of* M *if and only if* $T_\tau(M) \in \mathscr{F}$ *and for each R–linear mapping* $\phi : X \to M$, *where* X *is a* τ*–torsionfree injective* R*–module such that* ker ϕ *contains no nonzero* τ*–pure submodules of* X, *there is an* R*–linear mapping* $f : X \to T_\tau(M)$ *such that* $\varphi \circ f = \phi$.

Proof. If $\varphi : T_\tau(M) \to M$ is a torsionfree precover, then the condition follows from the definition of a precover. Conversely, suppose that the conditions holds. Let $\phi : X \to M$ be an R–linear mapping where $X \in \mathscr{F}$. By Proposition 5.23 and Zorn's lemma, let P be maximal among the pure submodules of X contained in ker ϕ. Then there is an induced R–linear mapping $\phi' : X/P \to M$ such that $\phi = \phi' \circ \eta$ where $\eta : X \to X/P$ is the canonical surjection. Moreover, ker ϕ' contains no nonzero pure submodules of X/P. Since M is an injective R–module, ϕ' can be extended to $\phi'' : E(X') \to M$ where $X' = X/P$ and $E(X')$ is the injective hull of X'. Note that $E(X') \in \mathcal{F}$ so that $E(X')$ is torsionfree and injective. If $\ker \phi''$ contains a pure submodule H of $E(X')$, then $(X' + H)/H \subseteq E(X')/H \in \mathcal{F}$. Thus, $X'/(X' \cap H) \in \mathcal{F}$ and $X' \cap H \subseteq \ker \phi'' \cap X' = \ker \phi'$. Hence, $X' \cap H = 0$ and so $H = 0$ since X' is essential in $E(X')$. Now, by assumption, there is an R–linear mapping $f' : E(X') \to T_\tau(M)$ such that $\varphi \circ f' = \phi''$. Thus, $\phi = \phi' \circ \eta = \phi'' \circ \eta = \varphi \circ f' \circ \eta$ and so if we let $f = f' \circ \eta$, we conclude that $\varphi : T_\tau(M) \to M$ is a τ–precover. □

The following proposition gives a sufficient condition for a faithful torsion theory on Mod–R to be universally covering. Conditions which are both necessary and sufficient are apparently yet to be discovered.

Proposition 5.3.6 *If* τ *is a faithful torsion theory on* Mod–R *and* $F_\tau(R)$ *contains a cofinal subset of finitely generated right ideals of* R, *then* τ *is universally covering.*

Proof. We will first show that every injective R–module M has a τ–torsionfree precover. Suppose that M is an injective R–module and let $\{ E_\alpha \}_{\alpha \in \Delta}$ be a set of representatives of the isomorphism classes of the τ–torsionfree cyclic R–modules. For each $\alpha \in \Delta$ and each $g \in \bar{\Delta} = \mathrm{Hom}_R(E_\alpha, M)$, let $E_{\alpha g}$ be a copy of E_α. Set $X = \oplus_{\alpha \in \Delta}(\oplus_{g \in \bar{\Delta}} E_{\alpha g})$. For each $k \in \Delta^* = \mathrm{Hom}_R(E(X), M)$, let X_k be a copy of $E(X)$. Next, let $M^* = \oplus_{k \in \Delta^*} X_k$ and define $\varphi^* : M^* \to M$ by $\varphi^*|_{X_k} = k$ for each k. We claim that $\varphi^* : M^* \to M$ is a τ–torsionfree precover of M. Note that φ^* is an epimorphism since τ if faithful and $E_\alpha \cong R$ for some $\alpha \in \Delta$. Since $M^* \in \mathcal{F}$,

Proposition 5.3.5 implies that $\varphi^* : M^* \to M$ is a τ–torsionfree precover of M if for any nonzero τ–torsionfree injective R–module Y and any $\phi : Y \to M$ such that $\ker \phi$ contains no nonzero τ–pure submodules of Y, there is an R–linear mapping $f : Y \to M^*$ such that $\varphi^* \circ f = \phi$. Let $\oplus_{\alpha \in B} Y_\alpha$ be a direct sum of injective hulls of nonzero cyclic R–modules such that $\oplus_{\alpha \in B} Y_\alpha$ is essential in Y. For $\alpha, \beta \in B$, define $Y_\alpha \sim Y_\beta$ if and only if there is an isomorphism $\pi_{\alpha\beta} : Y_\alpha \to Y_\beta$ such that $\phi|_{Y_\alpha} = (\phi|_{Y_\beta}) \circ \pi_{\alpha\beta}$. Then $\sim$ is an equivalence relation on $\{ Y_\alpha)_{\alpha \in B}$. If $\alpha \neq \beta$ and $Y_\alpha \sim Y_\beta$, define $G_{\alpha\beta} = \{ x - \pi_{\alpha\beta}(x) \mid x \in Y_\alpha \}$. Then $G_{\alpha\beta}$ is a submodule of Y and $\theta : G_{\alpha\beta} \to Y_\beta : x - \pi_{\alpha\beta}(x) \to \pi_{\alpha\beta}(x)$ is an isomorphism. But for $x - \pi_{\alpha\beta}(x) \in G_{\alpha\beta}$, $(\phi|_{G_{\alpha\beta}})(x - \pi_{\alpha\beta}(x)) = \phi(x) - \phi(\pi_{\alpha\beta}(x)) = 0$ by the definition of $\sim$. Thus, $G_{\alpha\beta}$ is a τ–pure and injective submodule of Y contained in $\ker \phi$, which contradicts our assumption. Therefore, each equivalence class determined by $\sim$ contains exactly one member.

For each $\alpha \in B$, there is an isomorphism $\theta_{\alpha\beta} : Y_\alpha \to E_\beta$ for some $\beta \in \Delta$. Define $f' = \oplus\, \theta_{\alpha\beta} : \oplus_{\alpha \in B} Y_\alpha \to X$ via $\theta_{\alpha\beta} : Y_\alpha \to E_{\beta(\phi\theta_{\alpha\beta}^{-1})}$. Then f' is a monomorphism since each equivalence class determined by $\sim$ contains only one member. Extend f' to $f'' : Y \to E(X)$ by the injectivity of $E(X)$. Since $\oplus_{\alpha \in B} Y_\alpha$ is essential in Y, f'' is a monomorphism. By the injectivity of M, choose a homomorphism $g : E(X) \to M$ such that $g \circ f'' = \phi$. Finally, define $f : Y \to M$ via $f = j \circ f''$, where j is the natural injection of X_g into M^*. Then $\varphi^* \circ f(x) = \varphi^*(f(x)) = \varphi^*(j(f''(x))) = g \circ f''(x) = \phi(x)$ for all $x \in Y$. Hence, $\varphi^* \circ f = \phi$. Hence, $\varphi^* : M^* \to M$ is a τ–torsionfree precover of M.

We have shown that if $F_\tau(R)$ contains a cofinal subset of finitely generated right ideals of R, then every injective R–module has a τ–torsionfree precover. By 1 of Proposition 5.3.4, if $\varphi^*: M^* \to E(M)$ is a τ–torsionfree precover of the injective hull of M, then

$\varphi^*|_{\varphi^{*-1}(M)} : \varphi^{*-1}(M^*) \to M$ is a τ–torsionfree precover of M. Hence, if every injective R–module has a τ–torsionfree precover, then every R–module has a τ–torsionfree precover. Now let $\varphi^* : M^* \to M$ be a τ–torsionfree precover of an arbitrary R–module M. By proposition 5.3.3 the union of an ascending chain of τ–pure submodule of M^* is a τ–pure submodule of M^*. Hence, use Zorn's Lemma to choose P maximal among the pure submodule of M^* which are contained in $\ker \varphi^*$. Then $\varphi : T_\tau(M) = M^*/P \to M : x + P \to \varphi^*(x)$ is a τ–torsionfree cover of M. □

Corollary 5.3.7 *If* τ *is a faithful torsion theory on* Mod–R *and* R *is* τ*–noetherian, then* τ *is universally covering.*

§6 CLASSICAL PROPERTIES OF RINGS RELATIVE TO A TORSION THEORY

6.1 (Semi)Primitive, Simple and (Semi)Prime Rings Relative to a Torsion Theory

The purpose of this section is to begin the study of the relation between simple, semiprime, prime, primitive and semiprimitive rings relative to a torsion theory on Mod–R.

Definition 6.1.1 An ideal K of a ring R is said to be a (right) τ*–primitive ideal* of R if K is the right annihilator of a cyclic τ–simple R–module. The ring R is said to be a (right) τ*–primitive ring* if 0 is a τ–primitive ideal of R. We will refer to R as being a τ*–semiprimitive ring* if R is τ–radical free. That is, if R is such that $J_\tau(R) = 0$.

It is implicit in the statement $J_\tau(R) = 0$ that R has τ–maximal right ideals. *Throughout this section we continue to assume that* τ *is a torsion theory on* Mod–R *such that* τ*–simple* R*–modules exist.*

Proposition 6.1.2 If M *is a* τ*–simple* R*–module, then* $(0 : M)$ *is a* τ*–pure ideal of* R.

Proof. If M is a τ–simple R–module and $m \in M$, then the R–linear mapping $R/(0:m) \to mR : r + (0:m) \to mr$ is an isomorphism. Hence, $(0:m) \in \mathscr{P}_\tau(R)$ and so the proposition follows from Proposition 2.1.4 since $(0:M) = \cap_{m \in M}(0:m)$. □

Corollary 6.1.3 *If* K *is a* τ*–primitive ideal of* R, *then* $K \in \mathscr{P}_\tau(R)$.

Corollary 6.1.4 *If* R *is a* τ*–primitive ring, then* $R \in \mathscr{F}$.

If K is a τ–primitive ideal of R, then R/K is a $\pi(\tau)$–primitive ring where $\pi(\tau)$ is the torsion theory on Mod–R/K induced on R/K by τ via the natural mapping $\pi : R \to R/K$. See Proposition 2.2.10 for details.

Proposition 6.1.5 *The following are equivalent:*

1. K *is a* τ*–primitive ideal of* R.
2. $K = (J : R)$ *for some* τ*–maximal right ideal* J *of* R.
3. K *is the largest ideal contained in some* τ*–maximal right ideal of* R.

Proof. $1 \Rightarrow 2$. Let K be the annihilator of the cyclic τ–simple R–module R/J. Then J is a τ–maximal right ideal of R and $r \in K$ if and only if $(R/J)r = 0$ if and only if $Rr \subseteq J$ if and only if $r \in (J : R)$. Thus, $K = (J : R)$.

$2 \Rightarrow 3$. Let $K = (J : R)$ where J is a τ–maximal right ideal of R and suppose $r \in K$. Then $r \in Rr \subseteq J$ and so $K \subseteq J$. Now suppose I is an ideal of R such that $K \subseteq I \subseteq J$. If $r \in I$, then $Rr \subseteq I \subseteq J$ and so $r \in (J : R) = K$. Hence, $K = I$.

$3 \Rightarrow 1$. Suppose K is the largest ideal contained in the τ–maximal right ideal J of R. Then R/J is a τ–simple R–module. Let A be the annihilator of R/J. If $r \in A$, then $r \in Rr \subseteq J$. Thus, $A \subseteq J$. Next, if $r \in K$, $Rr \subseteq K \subseteq J$ and so $(R/J)r = 0$. Hence, $r \in A$ and therefore $K \subseteq A$. Thus, $K = A$. □

Proposition 6.1.6 R *is a* τ*–primitive ring if and only if* R *admits a faithful cyclic* τ*–simple* R*–module.*

Proof. Immediate from the definition of a τ–primitive ring. □

Proposition 6.1.7 *The following ideals of* R *are equal.*

1. $J_\tau(R)$
2. $J_1 = \cap_{M \in \mathscr{C}} (0 : M)$ *where* $\mathscr{C}$ *is the class of* τ*–simple* R*–modules.*
3. $J_2 = \cap_{M \in \mathscr{P}} I$ *where* $\mathscr{P}$ *is the set of* τ*–primitive ideals of* R.

Proof. That $J_\tau(R) = J_1$ is Proposition 2.2.8. If $I \in \mathscr{P}$, then $I = (0 : M)$ for some cyclic τ–simple R–module M. Hence, it follows that $J_1 = \cap_{M \in \mathscr{C}} (0 : M) \subseteq \cap_{I \in \mathscr{P}} I = J_2$. To complete the proof observe that $MJ_2 = 0$ for every cyclic τ–simple R–module M. Hence, if K is a τ–maximal right ideal of R, then $(R/K)J_2 = 0$ and so $RJ_2 \subseteq K$. Hence, $J_2 \subseteq K$. Therefore, J_2 is contained in every τ–maximal right ideal of R and so $J_2 \subseteq J_\tau(R) = J_1$. □

Proposition 6.1.8 R *is a* τ*–semiprimitive ring if and only if* R *admits a faithful* τ*–radical free* R*–module* M.

Proof. Suppose that a faithful τ–radical free R–module M exists. By Proposition 2.2.6, $MJ_\tau(R) \subseteq J_\tau(M) = 0$. But M is faithful, so $J_\tau(R) = 0$. Hence, R is τ–semiprimitive. Conversely, if R is τ–semiprimitive, form $\{ M_\alpha \}_{\alpha \in \Delta}$ by choosing a representative from each of the isomorphism classes of τ–simple R–modules. Then, by Proposition 6.1.7, $M = \oplus_{\alpha \in \Delta} M_\alpha$ is such that $(0 : M) = \cap_{\alpha \in \Delta} (0 : M_\alpha) = J_\tau(R) = 0$. Next, note that 0 is a maximal

τ–pure submodule of each M_α since each M_α is τ–simple. Thus, $J_\tau(M_\alpha) = 0$ for each $\alpha \in \Delta$. But Proposition 2.2.11 shows that $J_\tau(M) = J_\tau(\oplus_{\alpha \in \Delta} M_\alpha) = \oplus_{\alpha \in \Delta} J_\tau(M_\alpha) = 0$ and so M is faithful and τ–radical free. □

Corollary 6.1.9 *If* R *is a* τ*–primitive ring, then* R *is* τ*–semiprimitive.*

Definition 6.1.10 A ring R is said to a *subdirect product* of a family $\{R_\alpha\}_{\alpha \in \Delta}$ of rings if there is a ring monomorphism $\varphi : R \to \prod_{\alpha \in \Delta} R_\alpha$ such that if $\pi_\beta : \prod_{\alpha \in \Delta} R_\alpha \to R_\beta$ is the canonical projection, $\pi_\alpha \circ \varphi$ is an epimorphism for each $\alpha \in \Delta$.

Proposition 6.1.11 *There is a torsion theory* τ *on* Mod–R *such that* R *is a* τ*–semiprimitive ring if and only if there exists a family of rings* $\{R_\alpha\}_{\alpha \in \Delta}$ *and a family of torsion theories* $\{\tau_\alpha\}_{\alpha \in \Delta}$ *such that:*

1. *For each* $\alpha \in \Delta$, τ_α *is a torsion theory on* Mod–R_α.
2. *For each* $\alpha \in \Delta$, R_α *is a* τ_α*–primitive ring.*
3. R *is a subdirect product of the family* $\{R_\alpha\}_{\alpha \in \Delta}$.

Proof. Let τ be a torsion theory on Mod–R such that R is τ–semiprimitive and let $\{A_\alpha\}_{\alpha \in \Delta}$ be the family of τ–primitive ideals of R. Then, by Proposition 6.1.7, $J_\tau(R) = \cap_{\alpha \in \Delta} A_\alpha = 0$. If for each $\alpha \in \Delta$, J_α is a τ–maximal right ideal of R such that $A_\alpha = (J_\alpha : R)$, then R/J_α is a cyclic τ–simple R–module whose right annihilator is A_α. For $\alpha \in \Delta$, let $F(R/A_\alpha)$ denote the Gabriel filter $\{K/A_\alpha \mid K \supseteq A_\alpha, K \in F_\tau(R)\}$ and suppose that τ_α is the torsion theory on Mod–R/A_α determined by $F(R/A_\alpha)$. Then R/J_α is a faithful cyclic τ_α–simple R/A_α–module for each $\alpha \in \Delta$ and R is a subdirect product of the family $\{R/A_\alpha\}_{\alpha \in \Delta}$.

Conversely, suppose there is a family of $\{ R_\alpha \}_{\alpha \in \Delta}$ of rings and a family $\{ \tau_\alpha \}_{\alpha \in \Delta}$ of torsion theories such that for each $\alpha \in \Delta$, R_α is a τ_α–primitive ring. Suppose furthermore that R is a subdirect product of this family of rings. Let $\varphi : R \to \prod_{\alpha \in \Delta} R_\alpha$ be a ring monomorphism such that $\pi_\alpha \circ \varphi$ is an epimorphism for each $\alpha \in \Delta$ where for each $\beta \in \Delta$, $\pi_\beta : \prod_{\alpha \in \Delta} R_\alpha \to R_\beta$ is the canonical projection. If for each $\alpha \in \Delta$ we let $A_\alpha = \ker(\pi_\alpha \circ \varphi)$, then through isomorphism we can assume that $R_\alpha = R/A_\alpha$. Moreover, $\cap_{\alpha \in \Delta} A_\alpha = 0$ since φ is injective. Since R/A_α is τ_α–primitive for each $\alpha \in \Delta$, there is a faithful cyclic τ_α–simple R/A_α–module $(R/A_\alpha)/(J_\alpha/A_\alpha) \cong R/J_\alpha$ and where for each $\alpha \in \Delta$, J_α/A_α is a τ_α–maximal right ideal of R/A_α. Next, let $\mathscr{K}$ denote the set of all right ideals K of R such that $K \supseteq A_\alpha$ for some $\alpha \in \Delta$ and such that $K/A_\alpha \in F(R/A_\alpha)$. Suppose furthermore that $F_\tau(R)$ is the intersection of all the Gabriel filters in R which contain $\mathscr{K}$. Then $F_\tau(R)$ is a Gabriel filter in R which induces a torsion theory τ on Mod–R. Since $A_\alpha = (J_\alpha : R)$ for each $\alpha \in \Delta$, then each A_α will be a τ–primitive ideal of R if we can show that each J_α is a τ–maximal right ideal of R. Suppose that $J_\alpha \subset K \subset R$ where K is a τ–pure right ideal of R. Then K/A_α is a τ_α–pure right ideal of R/A_α. Hence, $R/K \cong (R/A_\alpha)/(K/A_\alpha)$ is a τ_α–torsionfree R/A_α–module. But $0 \neq K/J_\alpha \subseteq R/J_\alpha$ and so R/K is a τ_α–torsion R/A_α–module and so we have a contradiction. Hence, J_α is a τ–maximal right ideal of R for each $\alpha \in \Delta$ and so A_α is τ–primitive for each $\alpha \in \Delta$. If $\mathscr{P}$ is the set of all τ–primitive ideals of R, then $\{ A_\alpha \}_{\alpha \in \Delta} \subseteq \mathscr{P}$ and so $\cap_{K \in \mathscr{P}} K \subseteq \cap_{\alpha \in \Delta} A_\alpha = 0$. Hence, $J_\tau(R) = 0$ and so R is a τ–semiprimitive ring. $\square$

Definition 6.1.12 A τ–torsionfree ring R is said to be a τ–*simple ring* if 0 and R are the only τ–pure ideals of R.

When the ring R has τ–maximal right ideals, it follows that if R is a τ–simple ring, then R is τ–radical free and consequently is τ–semiprimitive.

Proposition 6.1.13 *Any τ–simple ring is τ–primitive.*

Proof. Suppose R is a τ–simple ring and let M be a τ–simple R–module. If $0 \neq m \in M$, then mR is τ–simple and so there is a τ–maximal right ideal K of R such that $R/K \cong mR$. Thus, R admits a cyclic τ–simple R–module R/K. If A is the annihilator of R/K, then, by Proposition 6.1.2, A is a τ–pure ideal of R. Hence $A = 0$ or $A = R$ since R is a τ–simple ring. Clearly, $A \neq R$ because $R/K \neq 0$. Hence, $A = 0$ and so R/K is a faithful R–module. The result now follows from Proposition 6.1.6. □

Definition 6.1.14 An ideal A of R is said to be *completely τ–pure* if for each R–module $M \in \mathcal{F}$, $M/MA \in \mathcal{F}$. A proper ideal K of R is said to be *τ–prime* if whenever A and B are completely τ–pure ideals of R such that $AB \subseteq K$, either $A \subseteq K$ or $B \subseteq K$. A proper ideal K of R is said to be *τ–semiprime* if $A^2 \subseteq K$ implies $A \subseteq K$ for each completely τ–pure ideal A of R. If 0 is a *τ–(semi)prime ideal* of R, then R is said to be a *τ–(semi)prime ring*.

Note that the set of completely τ–pure ideals of R is nonempty since 0 and R are completely τ–pure ideals of R. When τ is faithful, a completely τ–pure ideal of R is a τ–pure ideal of R. This follows for if τ is faithful and

A is a completely τ–ideal pure of R, then $R/A = R/RA \in \mathcal{F}$. If the torsion theory τ is cohereditary, then every right ideal of R is completely τ–pure.

Obviously, any τ–prime ideal of R is τ–semiprime.

Proposition 6.1.15 *Any τ–(semi)primitive ring is τ–(semi)prime.*

Proof. If R is τ–semiprimitive, then, as in the proof of Proposition 6.1.8, R admits a faithful τ–radical free R–module of the form $M = \oplus_{\alpha \in \Delta} M_\alpha$ where each M_α is τ–simple. Now suppose that $A \neq 0$ is a completely τ–pure ideal of R. Since M is faithful, there must be at least one α such that $M_\alpha A \neq 0$ and so we can assume that $M_\alpha A \neq 0$ for all $\alpha \in \Delta$. A is completely τ–pure, so $M_\alpha / M_\alpha A \in \mathcal{F}$. But M_α is τ–simple which tells us that $M_\alpha / M_\alpha A \in \mathcal{T}$. Hence, it must be the case that $M_\alpha A = M_\alpha$ for each $\alpha \in \Delta$. Thus, $M_\alpha A^2 = M_\alpha A = M_\alpha$ for each $\alpha \in \Delta$ and so $\oplus_{\alpha \in \Delta}(M_\alpha A^2) = \oplus_{\alpha \in \Delta} M_\alpha = M$. Hence, $A^2 \neq 0$ and, consequently, $A^2 = 0$ implies $A = 0$. Therefore, R is τ–semiprime.

If R is a τ–primitive ring, then, by Proposition 6.1.6, R admits a faithful cyclic τ–simple R–module M. But then if A is a nonzero completely τ–pure ideal of R, using the same argument as above for M_α, $MA = M$. Hence, if B is also a nonzero completely τ–pure ideal of R, $M(AB) = (MA)B = MB = M$. Therefore, $AB \neq 0$. Thus, $AB = 0$ implies $A = 0$ or $B = 0$ and so 0 is a τ–pime ideal of R. □

Corollary 6.1.16 *Any τ–primitive ideal of* R *is τ–prime.*

Proof. Suppose that A is a τ–primitive ideal of R. Then, by Proposition 6.1.5, $A = (J : R)$ where J is a τ–maximal right ideal of R and R/J is a cyclic

τ–simple R–module. Next, let τ_A denote the torsion theory on Mod–R/A determined by the Gabriel filter $F_{\tau_A}(R/A) = \{ K/A \mid K \supseteq A, K \in F_\tau(R) \}$ (See Proposition 2.2.9). Then, by Proposition 6.1.6, R/J is a faithful cyclic τ_A–simple R/A–module and so R/A is a τ_A–primitive ring. Hence, by the proposition above, R/A is a τ_A–prime ring and from this it follows that A is a τ–prime ideal of R. □

We have shown the following ring implications which correspond directly to the implications for simple, semiprimitive, primitive, semiprime and prime rings.

		τ–semiprimitive	$\Rightarrow$	τ–semiprime
		$\Uparrow$		$\Uparrow$
τ–simple	$\Rightarrow$	τ–primitive	$\Rightarrow$	τ–prime

If R is a τ–primitive τ–artinian ring, then R a is τ–simple ring. To see this, let K be a τ–pure ideal of R. Since R/K is τ–artinian as an R–module, R/K has, by Proposition 2.4.3, a minimal nonzero τ–pure submodule N/K which is, by Proposition 2.1.8, τ–simple. Let $0 \neq x + K \in N/K$. Then $(x + K)R$ is a cyclic τ–simple R–module and so there is a τ–maximal right ideal J of R such that $R/J \cong (x + K)R$. If $A = (J : R)$, A is, by Proposition 6.1.5, a τ–primitive ideal of R. Thus, $A = 0$. But $K \subseteq A$ and so R is a τ–simple ring. Hence, we have the following proposition.

Proposition 6.1.17 *If* R *is a* τ*–primitive* τ*–artinian ring, then* R *is a* τ–simple *ring.*

We do not know what other implications given in the table above reverse under the assumption that R is τ–artinian.

6.2 Semisimple Rings Relative to a Torsion Theory

Recall that an R–module is semisimple if it is the sum of its simple submodules. This observation is the motivation for the following definition.

Definition 6.2.1 An R–module M is said to be τ–*semisimple* provided that M has at least one τ–simple submodule and if M is the sum of its τ–simple submodules. A ring R is said to be a τ–*semisimple ring* if R is τ–semisimple as an R–module.

Proposition 6.2.2 *If* $M = \sum_{\alpha \in \Delta} M_\alpha$ *is a* τ*–semisimple* R*–module where* M_α *is a* τ*–simple submodule of* M *for each* $\alpha \in \Delta$, *then there is a subset* Ω *of* Δ *such that the sum* $\sum_{\alpha \in \Omega} M_\alpha$ *is direct and* τ*–dense in* M.

Proof. Let $\mathscr{S}$ denote the collection of subsets Λ of Δ such that the sum $\sum_{\alpha \in \Lambda} M_\alpha$ is direct. Partial order $\mathscr{S}$ by inclusion and note that $\mathscr{S}$ is nonempty since subsets of Δ containing a single element are in $\mathscr{S}$. If $\{\Lambda_k\}_{k \geq 1}$ is a chain in $\mathscr{S}$, we claim that $\Lambda = \cup_{k \geq 1} \Lambda_k$ is in $\mathscr{S}$. If $0 = \sum_{\alpha \in \Lambda} m_\alpha \in \sum_{\alpha \in \Lambda} M_\alpha$, then $m_\alpha = 0$ for all but possibly a finite number of $\alpha \in \Delta$. Form the finite subset Γ of Λ by choosing the subscripts where m_α might not be zero. If k is such that $\Gamma \subseteq \Lambda_k$, then the sum $\sum_{\alpha \in \Gamma} M_\alpha$ is direct. Since $0 = \sum_{\alpha \in \Lambda} m_\alpha \in \sum_{\alpha \in \Gamma} M_\alpha$, it follows that $m_\alpha = 0$ for all $\alpha \in \Gamma$ and so $m_\alpha = 0$ for all $\alpha \in \Lambda$. Hence, $\sum_{\alpha \in \Lambda} M_\alpha$ is direct and so $\Lambda \in \mathscr{S}$. Consequently, by Zorn's Lemma, $\mathscr{S}$ has a maximal element Ω. We claim that $\sum_{\alpha \in \Omega} M_\alpha$ is τ–dense in R. Let $N = \sum_{\alpha \in \Omega} M_\alpha$ and note that $N \cap M_\alpha \neq 0$ for each $\alpha \in \Delta$. Thus, $M_\alpha/(N \cap M_\alpha)$ is τ–torsion for all $\alpha \in \Delta$

since each M_α is τ–simple. Hence, $\oplus_{\alpha \in \Delta} M_\alpha/(N \cap M_\alpha)$ is τ–torsion. Now $\oplus_{\alpha \in \Delta} M_\alpha/(N \cap M_\alpha) \to M/N : (m_\alpha + (N \cap M_\alpha)) \to \left(\sum_{\alpha \in \Lambda} m_\alpha\right) + N$ is an R–linear epimorphism and so M/N is τ–torsion. □

Proposition 6.2.3 *Any* R*–linear mapping from a* τ*–simple* R*–module to a* τ*–torsionfree* R*–module is either zero or a monomorphism.*

Proof. Suppose N is τ–simple and M is τ–torsionfree. If $f : N \to M$ is a nonzero R–linear mapping and $\ker f \neq 0$, then $N/\ker f$ is isomorphic to a submodule of M and so is τ–torsionfree. But $N/\ker f$ is τ–torsion since N is τ–simple. Hence, $N/\ker f \in \mathcal{T} \cap \mathcal{F} = 0$ and so $\ker f = N$ which contradicts the assumption that $f \neq 0$. Thus, when $f \neq 0$, $\ker f = 0$. □

Proposition 6.2.4 *The endomorphism ring* E_Q *of the quasi–injective hull of a* τ*–simple* R*–module* M *is a division ring. Furthermore, the endomorphism ring* E *of* M *embeds in* E_Q .

Proof. Let M be a τ–simple R–module and consider $E_\tau(M)$. We claim that $E_\tau(M)$ is τ–simple. First, note that $0 = t_\tau(M) = M \cap t_\tau(E_\tau(M))$ and so $E_\tau(M)$ is τ–torsionfree since M is essential in $E_\tau(M)$. Next, suppose that N is a nonzero submodule of $E_\tau(M)$. Then $N \cap M \neq 0$ and so the sequence $0 \to M/(N \cap M) \cong (M + N)/N \to E_\tau(M)/N \to E_\tau(M)/(M + N) \to 0$ is exact. Since M is τ–simple, $M/(N \cap M)$ is τ–torsion. Likewise, $E_\tau(M)/(M + N)$ is τ–torsion since it is a homomorphic image of $E_\tau(M)/M$ which is τ–torsion. Hence, $E_\tau(M)/N$ is τ–torsion since the class of τ–torsion R–modules is closed under extensions. Thus, $E_\tau(M)$ is a τ–simple R–module and so every nonzero submodule of $E_\tau(M)$ is τ–dense in $E_\tau(M)$. From this it follows that

$E_\tau(M)$ is quasi–injective and so contains the quasi–injective hull [26] $Q(M)$ of M. Notice that $Q(M)$ is also τ–simple since nonzero submodules of τ–simple R–modules are τ–simple. We claim that $E_Q = \mathrm{End}_R(Q(M))$ is a division ring. To see this, suppose $0 \neq f \in E_Q$. Then, by Proposition 6.2.3, f is injective and so there is an R–linear mapping $g : f(Q(M)) \to Q(M)$ such that $g \circ f = 1_{Q(M)}$. Since $Q(M)$ is quasi–injective, g extends to a map $h \in E_Q$. Consequently, every element of E_Q has a left inverse in E_Q and this is sufficient to show that E_Q is a division ring. □

Finally, suppose $f \in E = \mathrm{End}_R(M)$. Then f can be extended to a map $g \in E_Q$. We claim that this extension is unique. If $h \in E_Q$ is another such extension of f, then $g - h$ is the zero map on M. Thus, $g - h$ induces an R–linear mapping from $Q(M)/M$ to $Q(M)$. But $Q(M)/M$ is τ–torsion and $Q(M)$ is τ–torsionfree and so this map must be the zero map. Thus, $g = h$. If g_f denotes the unique extension of $f \in E$ to $Q(M)$, it follows that $E \to E_Q : f \to g_f$ is an embedding. □

Corollary 6.2.5 *The endomorphism ring of a quasi–injective τ–simple R–module is a division ring and the endomorphism ring of a τ–simple R–module is a domain.*

Proposition 6.2.6 *If M and N are τ–simple R–modules, then any nonzero map in $\mathrm{Hom}_R(E_\tau(M), E_\tau(N))$ is an isomorphism.*

Proof. Let $0 \neq f \in \mathrm{Hom}_R(E_\tau(M), E_\tau(N))$. As we have seen in the proof of Proposition 6.2.4, both $E_\tau(M)$ and $E_\tau(N)$ are τ–simple. Consequently, by Proposition 6.2.3, f is a monomorphism. Next, note that the short exact sequence $0 \to f(E_\tau(M)) \to E_\tau(N) \to E_\tau(N)/f(E_\tau(M)) \to 0$ splits since $f(E_\tau(M))$ is τ-injective and $E_\tau(N)/f(E_\tau(M))$ is τ-torsion. Thus, $f(E_\tau(M))$ is a direct summand of $E_\tau(N)$. But Proposition 3.1.10 shows that $f(E_\tau(M))$ is essential in $E_\tau(N)$ since $f(E_\tau(M))$ is τ-dense in $E_\tau(N)$. Thus, $f(E_\tau(M)) = E_\tau(N)$. □

Proposition 6.2.7 *If* τ *is a faithful torsion theory on* Mod–R, *then the following hold for any* τ*-semisimple* τ*-artinian ring* R.

1. *There exists a finite set* $\{K_i\}_{i=1}^n$ *of* τ*-simple right ideals of* R *whose sum is direct and* $\oplus_{i=1}^n K_i$ *is* τ*-dense in* R.
2. R *embeds as a ring in a direct product of matrix rings* $M_{k_1}(D_1) \times M_{k_2}(D_2) \times \cdots \times M_{k_q}(D_q)$ *where, for each* i, D_i *is the division ring* $\mathrm{End}_R(H_i)$ and $\{H_1, H_2, \ldots, H_q\}$ *is a complete set of representatives of the homogeneous components of* $\oplus_{i=1}^n E_\tau(K_i)$. *Furthermore,* R *is* τ*-dense in* $M_{k_1}(D_1) \times M_{k_2}(D_2) \times \cdots \times M_{k_q}(D_q)$.

Proof. 1. If R is τ-semisimple, then, by Proposition 6.2.2, there is a set $\{K_\alpha\}_{\alpha \in \Omega}$ of τ-simple right ideals of R whose sum is direct and such that $\oplus_{\alpha \in \Omega} K_\alpha$ is τ-dense in R. Since R is τ-artinian, Proposition 3.1.12, shows that R is also τ-noetherian. Suppose that the set Ω is not finite. Using the family $\{K_\alpha\}_{\alpha \in \Omega}$, we can construct an ascending chain $K_1 \subseteq K_1 \oplus K_2 \subseteq \cdots \subseteq K_1 \oplus K_2 \oplus \cdots \oplus K_k \subseteq \cdots$. It follows from Proposition 2.3.3 that $(K_1 \oplus K_2 \oplus \cdots \oplus K_{k+1})/(K_1 \oplus K_2 \oplus \cdots \oplus K_k)$ is τ-torsion for all $k \geq n$ for some positive integer n. But $(K_1 \oplus K_2 \oplus \cdots \oplus K_{k+1})/(K_1 \oplus K_2 \oplus \cdots \oplus K_k) \cong K_{k+1}$ is τ-torsionfree. Hence, it must be the case that $K_{k+1} = 0$ for all $k \geq n$ which contradicts the fact that each K_α must be nonzero to be τ-simple. Hence, Ω is a finite set.

2. Suppose $\Omega = \{1, 2, \ldots, n\}$. Since R is τ-torsionfree,

Proposition 3.1.10 indicates that $\oplus_{i=1}^{n} K_i$ is essential in R. Therefore, there is an R–linear embedding of R into $E_\tau(\oplus_{i=1}^{n} K_i) \cong E_\tau\left(\prod_{i=1}^{n} K_i\right) \cong \prod_{i=1}^{n} E_\tau(K_i)$. If $f \in End_R(R)$, then f extends uniquely to a $g \in End_R\left(\prod_{i=1}^{n} E_\tau(K_i)\right)$ and so we have a ring monomorphism $R \cong End_R(R) \to End_R\left(\prod_{i=1}^{n} E_\tau(K_i)\right)$. Suppose $\prod_{i=1}^{n} E_\tau(K_i) \cong \prod_{i=1}^{m} (E_\tau(K_{i1}) \times E_\tau(K_{i2}) \times \cdots \times E_\tau(K_{ik_i}))$ where thegroupings represent the rearrangement of the factors of $\prod_{i=1}^{n} E_\tau(K_i)$ into isomorphic collections and $k_1 + k_2 + \cdots + k_q = n$. For $i = 1, 2, \ldots, q$, choose a representative H_i from each homogeneous component of $\prod_{i=1}^{n} E_\tau(K_i)$ and note that $E_\tau(K_{i1}) \times E_\tau(K_{i2}) \times \cdots \times E_\tau(K_{ik_i}) \cong H_i^{k_i}$ for $i = 1, 2, \ldots, q$. Hence, $End_R\left(\prod_{i=1}^{n} E_\tau(K_i)\right) \cong End_R\left(\prod_{i=1}^{q} H_i^{k_i}\right)$ and each matrix $\mathcal{M}$ of the $n \times n$ matrix ring $End_R\left(\prod_{i=1}^{q} H_i^{k_i}\right)$ can be represented as a block matrix

$$\begin{pmatrix} \mathcal{M}_{11} & \mathcal{M}_{12} & \cdot & \cdot & \cdot & \mathcal{M}_{1q} \\ \mathcal{M}_{21} & \mathcal{M}_{22} & \cdot & \cdot & \cdot & \mathcal{M}_{2q} \\ \cdot & \cdot & \cdot & \cdot & \cdot & \cdot \\ \cdot & \cdot & \cdot & \cdot & \cdot & \cdot \\ \cdot & \cdot & \cdot & \cdot & \cdot & \cdot \\ \mathcal{M}_{q1} & \mathcal{M}_{q2} & \cdot & \cdot & \cdot & \mathcal{M}_{qq} \end{pmatrix}.$$

For $i, j = 1, 2, \ldots, m$, $\mathcal{M}_{ij}$ is a $k_i \times k_j$ matrix from the matrices determined by $Hom_R(H_j^{k_j}, H_i^{k_i}) \cong \prod^{k_j} Hom_R(H_j, H_i^{k_i}) \cong \prod^{k_j}\prod^{k_i} Hom_R(H_j, H_i)$ where $\prod^{k_j} Hom_R(H_j, H_i^{k_i})$ represents the direct product of k_j factors of $Hom_R(H_j, H_i^{k_i})$, etc. Thus, if $i \neq j$, Proposition 6.2.6 shows that $\mathcal{M}_{ij} = 0$ since, in this case, H_j is not isomorphic to H_i. If $i = j$, $\mathcal{M}_{ii}$ is a $k_i \times k_i$ matrix from the matrix ring $\mathcal{M}_{k_i}(End_R(H_i)) = Hom_R(H_i^{k_i}, H_i^{k_i})$ and Corollary 6.2.5 shows that $End_R(H_i)$ is a division ring D_i. Finally, observe that since $\prod_{i=1}^{n} K_i$ is τ–dense in $\prod_{i=1}^{n} E_\tau(K_i)$, R is τ–dense in $\prod_{i=1}^{n} E_\tau(K_i)$. □

Suppose that τ is the torsion theory in which every module is torsion–free. In this setting, the assumption in the proposition above reduces to R being a semisimple (artinian) ring. It follows that $\oplus_{i=1}^{n} K_i = R \cong M_{k_1}(D_1) \times M_{k_2}(D_2) \times \cdots \times M_{k_q}(D_q)$. In addition, $\{ H_1, H_2, \dots, H_q \}$ is a complete set of representatives of the homogeneous components of the direct sum $\oplus_{i=1}^{n} K_i$ decomposition of R by the irreducible right ideals K_i of R.

Notice that in Proposition 6.2.7 it was assumed that R is τ–semisimple *and* τ–artinian. Of course in the torsion theory τ in which every module is torsionfree, a τ–semisimple ring is τ–artinaian. One wonders as to what conditions are both necessary and sufficient for a τ–semisimple ring to be τ–artinian.

6.3 Primitive Rings and Density Relative to a Torsion Theory

Let R be a subring of the ring E of endomorphism of a vector space ${}_{\Delta}M$ where Δ is a division ring. R is said to be dense in E if for any $f \in E$ and any finite dimensional subspace U of V there exists an $r \in R$ such that $r|_U = f|_U$. The Jacobson Density Theorem [25] states that a ring R admits a faithful simple module if and only if R is (isomorphic to) a dense subring of the ring of endomorphism of a vectors space over a division ring. Such rings R are said to be primitive and they are characterized by the property that 0 is the largest ideal contained in some maximal right ideal of R. The density theorem might be viewed as saying that locally (= on finite dimensional subspaces) every endomorphism of ${}_{\Delta}V$ is a linear transformation in R.

The condition that for any $f \in E$ and any finite dimensional subspace U of V there exists an $r \in R$ such that $r|_U = f|_U$ can be replaced by the following condition: if $x_1, x_2, \dots, x_n$ and $y_1, y_2, \dots, y_n$ are vectors in V with the set $x_1, x_2, \dots, x_n$ being linearly independent in the left Δ–vector space V over the division ring Δ, then there is an $r \in R$ such that $x_i r = y_i$ for $i = 1, 2, \ldots, n$. This is the most common formulation of this condition. We now agree to write linear transformations on the side of the argument opposite that of scalars so that the ring opposite to R can be avoided when embedding R into E. To simplify notation we also eliminate the "o" used previously to indicate composition of functions and composition of functions written on the right of the argument will be from left to right.

The term density in the Jacobson Density Theorem has a different meaning than how the word is used in torsion theory. There is a topological version of the Jacobson Density Theorem in which a topology is introduced on E in such a way that R is topologically dense in E. For example, see [25].

When τ is a torsion theory on Mod–R, we have called a ring τ–primitive when R admits a faithful cyclic τ–simple R–module. This has been shown to be equivalent to 0 being the largest ideal contained in some τ–maximal right ideal of R. We conclude our study of selected topics in torsion theory by showing that a form of density also holds for τ–primitive rings. The following work is an adaption of several of the results of Zelmanowitz [39] to a torsion theoretical setting.

Proposition 6.3.1 *Let* R *be a subring of* $E = \text{End}_\Delta(M)$ *where* M *is a left* Δ*–vector space over a division ring* Δ*. Then the following are equivalent:*

1. *Given a linearly independent set* $\{ v_1, v_2, \ldots, v_n \}$ *of vectors in* M *and* $u_1, u_2, \ldots, u_n \in M$ *with* $u_1 \neq 0$*, there exists* $r, s \in R$ *with* $u_i r = v_i s$ *for* $i = 1, 2, \ldots, n$ *and* $u_1 r \neq 0$.
2. *Given a finite dimensional subspace* N *of* M *and* $f, g \in E$, $g \neq 0$*, there exists* $r, s \in R$ *such that* $fr|_N = s|_N$ *and* $gr \neq 0$.
3. *Given a finite dimensional subspace* N *of* M *and* $f \in E$ *with* $f|_N \neq 0$*, there exist* $r, s \in R$ *with* $fr|_N = s|_N \neq 0$.

Proof. $2 \Rightarrow 3$. If we choose $f \neq 0$ for g in 2, then there exist $r, s \in R$ such that $fr|_N = s|_N$ and $fr \neq 0$.

$3 \Rightarrow 1$. Let N be the subspace of M generated by $v_1, v_2, \dots, v_n$. Choose a basis $\mathscr{B}$ for M such that $\{ v_1, v_2, \dots, v_n \} \subseteq \mathscr{B}$ and define $f : \mathscr{B} \to M$ by $v_i f = u_i$ for $i = 1, 2, \dots, n$ and $vf = 0$ if $v \in \mathscr{B} \setminus \{ v_1, v_2, \dots, v_n \}$. Extend f linearly to E and note that $f|_N \neq 0$ since $v_1 f = u_1 \neq 0$. Hence, there exist $r, s \in R$ such that $fr|_N = s|_N \neq 0$ and so $v_i fr = v_i s$ for each i. Therefore, $u_i r = v_i s$ for $i = 1, 2, \dots, n$. If $u_1 r \neq 0$, we are finished, so suppose $u_1 r = v_1 s = 0$. Next, define $g : \mathscr{B} \to M$ by $v_1 g = u_1$ and $vg = 0$ if $v \in \mathscr{B} \setminus \{ v_1 \}$. Then $g|_N \neq 0$ and so there exist $r', s' \in R$ such that $gr'|_N = s'|_N \neq 0$. This yields $u_1 r' = v_1 s' \neq 0$ and $u_i r' = v_i s' = 0$ for $2 \leq i \leq n$. Hence, we see that $u_1(r + r') = u_1 r + u_1 r' = v_1 s' \neq 0$ and $u_i(r + r') = v_i(s + s')$ for $i = 1, 2, \dots, n$.

$1 \Rightarrow 2$. Let $f, g \in E$ with $g \neq 0$. There are two cases to be considered:

Case I. $(N)f \cap (M)g \neq 0$. Let $v_1 \in N$ and $u \in M$ be such that $v_1 f = ug \neq 0$. Extend v_1 to a basis $v_1, v_2, \dots, v_n$ to a basis for N. By hypothesis, there exist $r, s \in R$ such that $v_i fr = v_i s$ for each i and $v_1 fr \neq 0$. Hence, $fr|_N = s|_N$ and $gr \neq 0$ since $ugr = v_1 fr \neq 0$.

Case II. $(N)f \cap (M)g = 0$. Choose a basis $v_1, v_2, \dots, v_k, \dots, v_n$ for N such that $v_1 f, v_2 f, \dots, v_k f$ is a basis for $(N)f$ and $v_{k+1}, \dots, v_n \in \ker f$. By hypothesis, there exist $r, s \in R$ such that $v_i fr = v_i s$ for $i = 1, 2, \dots, n$ and $v_1 fr \neq 0$. Hence, $fr|_N = s|_N$. If $gr \neq 0$, the proof is complete, so now suppose $gr = 0$. Pick a $v \in M$ such that $vg \neq 0$. Since $(N)f \cap (M)g = 0$, the set of vectors $vg, v_1 f, \dots, v_k f$ is linearly independent. Again, by hypothesis, there exist $r', s' \in R$ such that $0 \neq vgr' = vs'$ and $v_i fr' = 0s' = 0$ for $1 \leq i \leq k$. Hence, $vg(r + r') = vgr + vgr' = vgr' \neq 0$ and for

$1 \le i \le k$, $v_i f(r + r') = v_i fr + (v_i)fr' = v_i s + v_i s' = v_i s$. For $k + 1 \le i \le n$, $v_i f(r + r') = v_i fr + v_i fr' = 0$ since $v_i \in \ker f$ when $k + 1 \le i \le n$. But $0 = v_i fr = v_i s$ for $k + 1 \le i \le n$. Thus, $v_i f(r + r') = v_i s$ for $i = 1, 2, \ldots, n$. Consequently, $f(r + r')|_N = s|_N$ and $g(r + r') \ne 0$. The proof is complete. □

Corollary 6.3.2 *Let* R *be a subring of* $E = End_\Delta(M)$ *where* M *is a left* Δ*−vector space over a division ring* Δ. *If* M *is finite dimensional as a left* Δ*−vector space and any one of the equivalent conditions of the proposition above is satisfied, then* E *is a rational extension of* R *as an* R*−module.*

Proof. By definition, an R–module M is a rational extension of a submodule N if and only if whenever $x, y \in M$ with $y \ne 0$, there is an $r \in R$ such that $xr \in N$ and $yr \ne 0$. Hence, if M is finite dimensional as a left Δ–vector space and $f, g \in E, g \ne 0$, then there exist $r, s \in R$ such that $fr|_M = s|_M$ and $gr \ne 0$. The corollary follows from the observation that, in this case, $fr = s \in R$ and $gr \ne 0$. (See [14, 15, 16, 37] for additional information on rational extensions of modules.) □

Definition 6.3.3 If R is a subring of $E = End_\Delta(M)$ where M is a left Δ–vector space over the division ring Δ, then R is said to be m–*dense* in E if any one of the three equivalent conditions of Proposition 6.3.1 is satisfied.

In order to simplify notation, we make the following definition.

Definition 6.3.4 If M is an R–module and N is a subset of M, then $N^R = (0 : N)$ and if S is a subset of R, then $S^M = \{ m \in M \mid mS = 0 \}$.

Before m–density can be tied to τ–primitive rings we need the following proposition.

Proposition 6.3.5 *If* M *is a quasi–injective* R–*module, let* $\Delta = \mathrm{End}_R(M)$. *If* N *is any* Δ–*submodule of* ${}_\Delta M$ *such that* $N^{RM} = N$, *then:*

1. $(N + \Delta m)^{RM} = N + \Delta m$ *for all* $m \in M$.
2. *If* $x_1, x_2, \ldots, x_n \in M$, *then* $\left(\sum_{i=1}^{n} \Delta m_i\right)^{RM} = \sum_{i=1}^{n} \Delta m_i$.

Proof. Observe that $r \in (\Delta m)^R$ if and only if $\Delta mr = 0$ if and only if $mr = 0$ if and only if $r \in m^R$. Hence, $m^R = (\Delta m)^R$.

1. Obviously, $(N + \Delta m)^{RM} \supseteq N + \Delta m$ and so we need to show the reverse containment. First, note that $(N + \Delta m)^R = N^R \cap (\Delta m)^R$ and so we must show that $(N^R \cap (\Delta m)^R)^M \subseteq N + \Delta m$. Let $m' \in (N^R \cap (\Delta m)^R)^M$ and consider the mapping $f : mN^R \to m'N^R : mr \to m'r$. If $mr = 0$, then $r \in m^R = (\Delta m)^R$ and so $r \in N^R \cap (\Delta m)^R$. Since $m' \in (N^R \cap (\Delta m)^R)^M$, we have that $m'r = 0$ and so f is well defined. Since f is R–linear and M is quasi–injective, f can be extended to a $g \in \Delta$. Thus, $g(mr) = m'r$ for all $r \in N^R$. But then $(g(m) - m')N^R = 0$ and so $g(m) - m' \in N^{RM} = N$. Thus, $m' \in N + \Delta m$ and so $(N + \Delta m)^{RM} = N + \Delta m$.

2. If we choose $N = 0$ in 1, then $(\Delta m_1)^{RM} = \Delta m_1$ and so we have the case for $n = 1$. Now suppose that $\left(\sum_{i=1}^{k} \Delta m_i\right)^{RM} = \sum_{i=1}^{k} \Delta m_i$. Then $(N + \Delta m_{k+1})^{RM} = N + \Delta m_{k+1}$ again by 1 with $N = \sum_{i=1}^{k} \Delta m_i$. The result now follows by induction. $\square$

Corollary 6.3.6 *If* M *is a quasi–injective* R–*module,* $\Delta = \mathrm{End}_R(M)$ *and* $m_1, m_2, \ldots, m_n \in M$, *then* $m \in \sum_{i=1}^{n} \Delta m_i$ *if and only if* $m^R \supseteq \cap_{i=1}^{n} m_i^R$.

Proof. If $m \in \sum_{i=1}^{n} \Delta m_i$, the containment is obvious. Suppose $m^R \supseteq \cap_{i=1}^{n} m_i^R$ and note that $\left(\sum_{i=1}^{n} \Delta m_i\right)^R = \cap_{i=1}^{n}(\Delta m_i)^R = \cap_{i=1}^{n} m_i^R \subseteq m^R$. Thus, $m\left(\sum_{i=1}^{n} \Delta m_i\right)^R = 0$ and so $m \in \left(\sum_{i=1}^{n} \Delta m_i\right)^{RM} = \sum_{i=1}^{n} \Delta m_i$. □

We can now establish a connection between τ–primitive rings and rings which are m–dense in an endomorphism ring of a left vector space over a division ring.

Proposition 6.3.7 *Suppose that* R *is a* τ*–primitive ring with* (cyclic) *faithful* τ*–simple* R*–module* M. *Then:*

1. $\Delta = \mathrm{End}_R(\bar{M})$ *is a division ring where* $\bar{M}$ *is the quasi–injective hull of* M.
2. R *embeds as a ring in* $E = \mathrm{End}_\Delta(\bar{M})$.
3. R *is* m*–dense in* E.

Proof. 1. Since the quasi–injective hull of a τ–simple R–module is τ–simple, this is Corollary 6.2.5.

2. Because M is faithful, $\bar{M}$ is faithful and so the canonical mapping $\varphi : R \to E : r \to f_r$ where $f_r : \bar{M} \to \bar{M} : m \to mr$ is a ring monomorphism. We now identify R with its image in E.

3. Let $v_1, v_2, \ldots, v_n \in \bar{M}$ be Δ–linearly independent and suppose $u_1, u_2, \ldots, u_n \in \bar{M}$ with $u_1 \neq 0$. We use induction to show that we can find $r, s \in R$ such that $u_i r = v_i s$ for $i = 1, 2, \ldots, n$ with $u_1 r \neq 0$. Proposition 3.1.10 shows that any τ–dense submodule of a τ–torsionfree R–module is essential. Hence, $u_1 R \cap v_1 R \neq 0$ and so the case for $n = 1$ is immediate. Now suppose that whenever $v_1, v_2, \ldots, v_k \in \bar{M}$ are linearly independent and $u_1, u_2, \ldots, u_k \in \bar{M}$, $u_1 \neq 0$, we can find $r, s \in R$ such that $u_i r = v_i s$ for

$i = 1, 2, \ldots, k$ and $u_1 r \neq 0$. Finally, suppose $v_1, v_2, \ldots, v_{k+1}$ are linearly independent in $\bar{M}$ and $u_1, u_2, \ldots, u_{k+1} \in \bar{M}$ with $u_1 \neq 0$. We can apply the induction hypothesis to find $r', s' \in R$ such that $u_i r' = v_i s'$ for $i = 1, 2, \ldots, k$ and $u_1 r' \neq 0$. If $u_{k+1} r' = v_{k+1} s'$ we are finished, so suppose $u_{k+1} r' \neq v_{k+1} s'$ and let $w_{k+1} = u_{k+1} r' - v_{k+1} s'$. We claim the proof will be complete if we can find $r'', s'' \in R$ such that $v_i s'' = 0$ for $1 \leq i \leq k$ and $w_{k+1} r'' = v_{k+1} s''$ and $u_1 r' r'' \neq 0$. For then $u_i r' r'' = v_i s' r'' = v_i s' r'' + v_i s'' = v_i(s' r'' + s'')$ when $1 \leq i \leq k$ and $u_{k+1} r' r'' = (v_{k+1} s' + w_{k+1}) r'' = v_{k+1}(s' r'' + s'')$ and $u_1 r' r'' \neq 0$. Hence, we need only choose $r = r' r''$ and $s = s' r'' + s''$. We now show that such an r'' and s'' can always be found. There are two cases to be considered.

Case I. $u_1 r' \in \Delta w_{k+1}$. Let $u_1 r' \in \Delta w_{k+1}$ and suppose $u_1 r' = d w_{k+1}$ where $d \in \Delta$ must be nonzero since $u_1 r' \neq 0$. Set $A = \cap_{i=1}^{k} v_i^R$, then, by Corollary 6.3.6, $v_{k+1} A \neq 0$ since the vectors $v_1, v_2, \ldots, v_{k+1}$ are linearly independent. Therefore, $v_{k+1} A \cap w_{k+1} R \neq 0$. Choose $r'', s'' \in R$, $s'' \in A$, to be such that $0 \neq w_{k+1} r'' = v_{k+1} s''$. Then $v_i s'' = 0$ for $1 \leq i \leq k$ because $s'' \in A$ and $u_1 r' r'' = d w_{k+1} r'' \neq 0$ because $w_{k+1} r'' \neq 0$ and $d \in \Delta$ is an R–linear isomorphism.

Case II. $u_1 r' \notin \Delta w_{k+1}$. In this case, $(u_1 r')^R \not\subset (w_{k+1})^R$ and $(w_{k+1})^R \not\subset (u_1 r')^R$ again by Corollary 6.3.6. Hence, $v_{k+1} A \cap w_{k+1}(u_1 r')^R \neq 0$ where, as before, $A = \cap_{i=1}^{k} v_i^R$. Choose $a \in (u_1 r')^R$ and $s'' \in A$ to be such that $w_{k+1} a = v_{k+1} s'' \neq 0$. Next choose $b \in (w_{k+1})^R \setminus (u_1 r')^R$ and set $r'' = a + b$. Then $v_i s'' = 0$ for $i = 1, 2, \ldots, k$, $v_{k+1} s'' = w_{k+1} a = w_{k+1} a + w_{k+1} b = w_{k+1} r''$ and $u_1 r' r'' = u_1 r'(a + b) = u_1 r' b \neq 0$. $\square$

In passing, notice that in the proposition above if τ is chosen to be the torsion theory in which every R–module is torsionfree, then R is primitive and so R in dense in E in the sense of Jacobson. This follows since the faithful (cyclic) τ–simple R–module M is now a faithful simple R–module.

Proposition 6.3.8 Let ${}_{\Delta}M$ *be a left vector space over a division ring* Δ *and suppose that* R *a subring of* $E = End_{\Delta}(M)$. *If* M *is cyclic as an* R*–module and* R *is* m*–dense in* E, *there exists a torsion theory* τ *on* Mod–R *such that* M *is a* τ–simple R*–module and* R *is a* τ*–primitive ring.*

Proof. Let τ be the torsion theory cogenerated by E(M), the injective hull of M. Since M is τ–torsionfree, M will be a τ–simple R–module if for every nonzero R–submodule N of M, M/N is τ–torsion. To show this, we will first show that any $0 \neq f \in Hom_R(N, M)$ is an injective mapping. Let $u \in N$ be such that $f(u) \neq 0$. If v is an arbitrary nonzero element of N, we need to show $f(v) \neq 0$. There are two cases which must be considered:

Case I. $f(u) \in \Delta v$. Let $t \in E$ be a transformation such that $vt = u$. Choose $r, s \in R$ such that $0 \neq vtr = vs$. Then $ur = vs$. If $f(u) = dv$ where $d \in \Delta$, $d \neq 0$, and $f(v) = 0$, then $0 = f(v)s = f(vs) = f(ur) = f(u)r = dur$. Hence, $ur = 0$ which is a contradiction. Thus, $f(v) \neq 0$ when $f(u) \in \Delta v$.

Case II. $f(u) \notin \Delta v$. In this case, f(u) and v are linearly independent over Δ, so let $r, s \in R$ be such that $0 \neq f(u)r$ and $ur = vs$. Then $0 \neq f(u)r = f(ur) = f(vs) = f(v)s$ and so $f(v) \neq 0$ when $f(u) \notin \Delta v$.

Next, if M/N is not τ–torsion, then $Hom_R(M/N, E(M)) \neq 0$. If $0 \neq g \in Hom_R(M/N, E(M))$, then $g(M/N) \cap M \neq \emptyset$, so let N′ be such that $N'/N = g^{-1}(M)$. If $\pi : N' \to N'/N$ is the canonical surjection, $g \circ \pi : N' \to M$ is nonzero and R–linear but not injective, contradicting our previous observation. Thus, M/N must be τ–torsion and so M is τ–simple.

It follows that M is a faithful R–module for $Mr = 0$ indicates that r is the zero transformation in E. Consequently, M is a faithful cyclic τ–simple R–module and so R is τ–primitive. □

Note that we have assumed that M is a cyclic R–module in the proposition above. If this condition can be shown to be necessary, then R is a τ–primitive ring for some torsion theory τ on Mod–R if and only if R is m–dense in a ring of linear transformations over a vector space. On the other hand, for M to be cyclic is not required in the proof of Proposition 6.3.6. If we define R to be τ–primitive if R admits a faithful τ–simple R–module, then R is τ–primitive for some torsion theory τ on Mod–R if and only if R is m–dense in a ring of linear transformations on a vector space. But then the intrinsic characterization of R being τ–primitive if and only if 0 is the largest ideal contained in some τ–maximal right ideal of R seems to be lost.

REFERENCES

1. R. Baer, Abelian groups which are direct summands of every containing ableian group, Bull. Amer. Math. Soc. **46**(1940), 800–806.
2. B. Banaschewski, On coverings of modules, Math. Nachr. **31**(1966), 57–71
3. H. Bass, Finitistic dimension and a homological generalization of semiprimary rings, Trans. Amer. Math. Soc. **95**(1960), 466–488.
4. H. Bass, Algebraic K–Theory, W. A. Benjamin, Inc., New York 1968.
5. P. Bland, A note on quasi–divisible modules, Comm. in Algebra **18**(6) (1990), 1953–1959.
6. P. Bland, Perfect torsion theories, Proc. Amer. Math. Soc. **41**(1973), 349–355.
7. P. Bland, Quasi–codivisible covers, Bull. Australian Math. Soc. **33** (1986), 389–395.
8. P. Bland, Relatively flat modules, Bull. Australian Math. Soc. **13** (1975), 375–387.
9. T. Cheatham, Finite dimensional torsion free rings, Pacific J. Math. **39** (1971), 113–118.
10. S. E. Dickson, A torsion theory for abelian categories, Trans. Amer. Math. Soc. **121**(1966), 223–235.
11. B. Eckman and A. Schopf, Über injektive Moduln, Archiv. der Math. **4**(1953), 75–78.
12. E. Enochs, Torsion free covering modules, Proc. Amer. Math. Soc. **14** (1963), 884–886.
13. E. Enochs, Torsion free covering modules II, Arch. Math. (Basel) **22** (1971), 37–52.
14. G. D. Findlay and J. Lambek, A generalized ring of quotients I, Canad. Math. Bull. **1**(1958), 77–85.
15. G. D. Findlay and J. Lambek, A generalized ring of quotients II, Canad. Math. Bull. **1**(1958), 155–167.
16. C. Faith, Lectures on Injective Modules and Quotient Rings, Lecture Notes in Mathematics, Springer–Verlag, Berlin 1967.
17. K. Fuller and D. Hill, On quasi–projective modules via relatively projectivity, Arch. Math. **21**(1970), 369–373.
18. L. Fuchs, On quasi–injective modules, Annali della Scuola Norm. Sup. Pisa **23**(1969), 541–546.
19. P. Gabriel, Des categories abeliennes, Bull. Soc. Math. France **90**(1962), 323–448.

20. J. S. Golan, Torsion Theories, Longman Scientific and Technical with John Wiley and Sons Inc., New York 1986.
21. J. S. Golan and M. L. Teply, Torsion–free covers, Israel J. Math. **15**(1973), 237–256.
22. A. W. Goldie, Torsion free modules and rings, J. Algebra **1**(1964), 268–287.
23. G. Goldman, Elements of noncommutative arithmetic, J. Algebra **35**(1975), 308–341.
24. C. Hopkins, Rings with minimal condition for left ideals, Ann. Math. **40**(1939), 712–730.
25. N. Jacobson, Structure of Rings, Amer. Math. Soc. Colloq. Publ., vol. 37, Providence, R. I., 1964.
26. R. Johnson and E. Wong, Quasi–injective modules and irreducible rings, J. London Math. Soc. **36**(1961), 260–268.
27. J. Levitzki, On rings which satisfy the minimum condition for right–hand ideals, Compositio Math. **7**(1939), 214–222.
28. A. Koehler, Quasi–projective covers and direct sums, Proc. Amer. Math. Soc. **24**(1970), 655–658.
29. E. Matlis, Injective modules over noetherian rings, Pacific J. Math. **8**(1958), 511– 528.
30. R. W. Miller and M. L. Teply, The descending chain condition relative to a torsion theory, Pacific J. Math. **83**(1979), 207–219.
31. B. Pareigis, Radikale und kleine Moduln, Bayer. Akad. Wiss. Math.–Natur. Kl. Sitzungsber **11**(1966), 185–199.
32. T. Porter, The kernels of completion maps and a relative form of Nakayama's lemma, Journal of Algebra **85**(1983), 166–178.
33. K. Rangaswamy, Codivisible modules, Comm. Algebra **2**(1974), 475–489.
34. M. L. Teply, Flatness relative to a torsion theory, Comm. Algebra **6**(1978), 1037–1071.
35. M. L. Teply, Torsion free covers II, Israel J. Math. **23**(1976), 132–136.
36. M. L. Teply, Torsion free injective modules, Pacific J. Math. **28**(1969), 441–453.
37. B. Stenstrom, Rings of Quotients, Springer–Verlag, Berlin 1975.
38. L. Wu and J. Jans, On quasi–projectives, Illinois J. Math. **11**(1967), 439–448.
39. J. M. Zelmanowitz, Representations of rings with faithful monoform modules, Comm. Algebra **14**(1986), 1141–1169.

INDEX

τ–artinian module, 49
ascending chain condition, 39

Baer's Condition, 66
balanced torsion theory, 16

category has enough τ–projective
modules, 85
character module, 89
closed under extensions, 11
closed under homomorphic
images, 11
closed under direct products, 11
closed under direct sums, 11
closed under submodules, 11
cofinal subset, 75
τ–cofinitely generated, 49
cohereditary torsion theory, 12
completely τ–pure ideal, 140
τ–compeletely reducible module, 67
complimentary pair, 12
τ–composition series, 55

τ–dense extension, 19
τ–dense submodule, 19
density (topological), 149
descending chain condition, 49
τ–divisible module, 27

essential submodule, 15
τ–essential submodule, 19
τ–essential extension, 19

faithful torsion theory, 12
filter, 19
finite homomorphism, 68
τ–finite length, 55
finitely τ–radical free, 59
τ–finitely generated, 40
τ–flat module, 88
free homomorphism, 113
Fuchs Condition, 78

Gabriel filter, 20
Generalized Baer Condition, 66
Generalized Fuchs Condition, 79
Generalized Hopkins–Levitzki
Theorem, 62
Generalized Nakayama Lemma, 48
Goldie torsion theory, 26

hereditary torsion class, 12
hereditary torsion theory, 12

injective hull, 11, 14
injective module, 14
τ–injective hull, 102
τ–injective module, 65

Jacobson Density Theorem, 148

Lambek torsion theory, 26
Lambek dense, 26
lenght of a τ–composition series, 55

m–dense, 151
τ–Max, 46
τ–maximal submodule, 29
minimal homomorphism, 113
τ–minimal homomorphism, 81, 113

Nakayama's Lemma, 45
τ–noetherian, 39
nonsingular module, 26

orthogonal compliment, 12

perfect ring, 16, 101
τ–prime ideal, 140
τ–prime ring, 140

τ–primitive ideal, 135
τ–primitive ring, 135
projective cover, 11, 16, 101
τ–projective cover, 113
projective module, 16
τ–projective module, 81
τ–pure closure, 31
τ–pure ideal, 30
τ–pure right ideal, 30
τ–pure submodule, 30

τ–quasi–injective hull, 110
τ_f–quasi–injective module, 78
τ–quasi–injective module, 80
τ–quasi–projective cover, 121
τ–quasi–projective module, 81

τ–radical, 33
τ–radical free module, 33
τ–reduced module, 28
right singular ideal, 26
right perfect ring, 16, 101
right τ–primitive ideal, 135
right τ–primitive ring, 135
right T–nilpotent, 16, 101

τ–semiprime ideal, 140
τ–semiprime ring, 140
τ–semisimple module, 143
τ–semisimple ring, 143
τ–semiprimitive ring, 135
τ–simple module, 29
τ–simple ring, 140
singular ideal, 26
singular module, 26
singular submodule, 26
small submodule, 16
τ–small submodule, 34
stable torsion theory, 74
subdirect product, 138

T–nilpotent, 16, 101
τ–torsion ideal, 14
τ–torsion module, 12
τ–torsion submodule, 14
torsion theory, 11
torsion class, 12
torsion theory cogenerated by, 24
torsion theory generated by, 24
torsionfree class, 12
torsionfree factor property, 126
τ–torsionfree cover, 127
τ–torsionfree module, 12
τ–torsionfree percover, 127

universally covering, 127